T0043578

BEEKEEPING *for* BEGINNERS

BEEKEEPING for BEGINNERS

HOW TO RAISE YOUR FIRST BEE COLONIES

Amber Bradshaw

ROCKRIDGE
PRESS

Copyright © 2019 by Rockridge Press. Emeryville. California

No part of this publication may be reproduced. stored in a retrieval system, or transmitted in any form or by any means, electronic, mechanical, photocopying, recording, scanning, or otherwise, except as permitted under Sections 107 or 108 of the 1976 United States Copyright Act, without the prior written permission of the Publisher. Requests to the Publisher for permission should be addressed to the Permissions Department. Rockridge Press. 6005 Shellmound Street. Suite 175. Emeryville. CA 94608.

Limit of Liability/Disclaimer of Warranty: The Publisher and the author make no representations or warranties with respect to the accuracy or completeness of the contents of this work and specifically disclaim all warranties, including without limitation warranties of fitness for a particular purpose. No warranty may be created or extended by sales or promotional materials. The advice and strategies contained herein may not be suitable for every situation. This work is sold with the understanding that the Publisher is not engaged in rendering medical, legal, or other professional advice or services. If professional assistance is required, the services of a competent professional person should be sought. Neither the Publisher nor the author shall be liable for damages arising herefrom. The fact that an individual, organization, or website is referred to in this work as a citation and/ or potential source of further information does not mean that the author or the Publisher endorses the information the individual, organization, or website may provide or recommendations they/it may make. Further, readers should be aware that Internet
websites listed in this work may have changed or disappeared between when this work was written and when it is read.

For general information on our other products and services or to obtain technical support. please contact our Customer Care Department within the U.S. at (866) 744-2665, or outside the U.S. at (510) 253-0500.

Rockridge Press publishes its books in a variety of electronic and print formats. Some content that appears in print may not be available in electronic books, and vice versa.

TRADEMARKS: Rockridge Press and the Rockridge Press logo are trademarks or registered trademarks of Callisto Media Inc. and/or its affiliates, in the United States and other countries, and may not be used without written permission. All other trademarks are the property of their respective owners. Rockridge Press is not associated with any product or vendor mentioned in this book.
Interior and Cover Designer: Joshua Moore
Art Producer: Sara Feinstein
Editor: Salwa Jabado
Production Editor: Andrew Yackira
Illustrations: © 2019 Benlin Alexander
Photographs: © Shutterstock. cover. p. ii, x, 4, 47, 72. 106. 126: iStock, cover, p. viii. 17, 24, 59, 74, 92, 110. 135, 138, 141: Edward Fury/Stocksy. cover: Michela Ravasio/Stocksy. cover, p. 48; Paul Tessier/Stocksy. p. 27: Borislav Zhuykov/Stocksy. p. 77: Studiofena/ Stocksy. p. 119: Pixel Stories/Stocksy. p. 123: Creative Market. p. 150.

ISBN: Print 978-1-64152-486-5
eBook 978-1-64152-487-2

TO MY HUSBAND,
TIMMY, AND OUR THREE
CHILDREN, GAVIN,
MORGAN, AND LINDEN

CONTENTS

Introduction IX

Part One
ALL ABOUT BEES AND SETTING UP AN APIARY

1
SO, YOU WANT TO KEEP BEES?
2

The Bees' Knees: The Benefits of Beekeeping 3

The Business End of a Bee: The Challenges of Beekeeping 7

Is Beekeeping Legal? 9

Other Things to Consider 10

2
HONEYBEE 101
14

Her Highness, the Queen 15

The Workers: The Ladies of the Hive 17

The Drones: Flavor of the Month 20

The Buzzing Hive in Nature 22

The Bee Life Cycle 24

3
A HOME FOR YOUR BEES: CHOOSING A HIVE
28

Hive Requirements 29

What Type of Beekeeper Do You Want to Bee? 32

An Introduction to Beehives 34

The Langstroth Hive 35

Top Bar Hives 37

The Warre Hive 41

To Buy or DIY? 42

Must-Have Beekeeping Supplies 46

4
PLANNING YOUR COLONY
52

Which Bee Is Best for You? 53

Buying Your Bees 55

A Honeybee Garden 59

Part Two

THE FIRST YEAR OF YOUR NEW COLONY

5

BRINGING HOME THE BEES
64

Bee Prepared 65

How to Transport Bees 66

Welcome Home 68

Release the Bees! 71

Feeding Your New Bees 76

6

THE BEE INSPECTOR
80

Timing Is Everything 81

How Often Should You Inspect the Hive? 83

Getting Ready for a Hive Inspection 86

All About Your Smoker 87

The Inspection Procedure 89

Start Your Beekeeping Journal 95

Be Prepared for Bee Stings 97

7

THE BEEKEEPING YEAR
100

The First Two Months 101

Spring, the First Year 102

Spring, the Second Year 102

Summer 105

Fall 107

Winter 111

8

ALL ABOUT HONEY
114

First Things First: What Is Honey? 115

It Takes a Lot of Bees 116

The Honey Harvest 117

Honey Extraction Equipment 120

Harvesting Honey, Step by Step 121

What Can You Do with All That Beeswax and Honey? 126

9

KEEPING YOUR BEES HEALTHY AND YOUR COLONY PRODUCTIVE

128

▼

Why Do Bees Get Sick? **129**

Signs of Sickness **131**

The Most Common Pests and What to Do About Them **134**

Critters to Watch Out For **138**

10

THE FUTURE OF YOUR BEES

142

▼

Do You Need More than One Hive? **143**

Don't Fear the Swarm **144**

A Royal Coup D'état **145**

Keep Things Clean **147**

Are You Ready to Start Beekeeping? **149**

Beekeeping Resources **151**

Glossary **154**

References **159**

Index **162**

INTRODUCTION

MY BEEKEEPING JOURNEY didn't start with a desire to raise bees; it started with the desire to grow my own food. With rising food costs, kids with allergies, and a lack of fresh local produce, I decided to turn every square inch of our urban yard into a food forest. For years I struggled with producing enough food to really make any difference in our lives. Our plants would flower but wouldn't set fruit. It was in my third year of unsuccessful gardening when I finally noticed that we didn't have any pollinators. It wasn't just the pollinators that had performed a disappearing act—it seemed like many of the insects that I grew up seeing were missing in action too. Everything but those darn mosquitoes. I swear, nothing kills those buggers!

It suddenly became clear, if I wanted to grow food, I needed pollinators. Better yet, I could be the pollinator.

Becoming a beekeeper had been the furthest thing from my mind. I mean, I'm allergic to bee stings, for goodness' sake. For an entire year, I went out to my garden with my tiny paintbrush and transferred the pollen from plant to plant. Talk about labor-intensive. Did you know the female flower is not always open at the same time as the male flower? I didn't. Turns out that plants are fickle and moody, just like humans. Needless to say, I soon realized I needed plan B(ee).

The question is: Where does a person start? You can't buy a pack of bees at your local grocery store and, living in an urban area with lots of tourists, it was almost impossible to find local resources that would guide me in the right direction. I was about as clueless as one could be about bees.

When it came to bees, I knew two things: one, they made honey, and two, bee stings would give me hives. Oh, and three, they would die if they stung you, which I learned from watching a cartoon with my daughter.

I started buying books and looking online for beginner beekeeping information. I signed up for a beekeeping course at my local extension agency—which wasn't too local, at an hour-and-a-half drive each way. The problem was, even with the course, I felt like I was supposed to show up at the class already knowing everything there was to know about bees. All the information that was available to me seemed like it was in a foreign language or over my head. Everyone else in the course appeared to know all the bee terminology—they lost me at "propolis." I needed a "beekeeping 101 for the person who knows nothing." But I couldn't find it. Frustrated, I decided to dive right in. How hard could it be to raise a bunch of bugs, right? *Wrong.*

Like most of my knowledge, my beekeeping experience has come from years of mistakes—trial and error, with many failures and successes along the way. I wanted to write a book that would help other people like me—people who want to help the bee population, invite pollinators into their garden, raise bees, and have a fresh supply of honey, but know nothing about how to do it or even where to start.

So that's what I did.

In this book, I'll take you from the very first steps of beekeeping to the last. You will, hopefully, be able to raise your own successful colonies while having a fresh supply of honey and thousands of healthy pollinators. I will take the mystery out of beekeeping and offer simple alternatives, so you can be successful instead of frustrated with your beekeeping experience. Beekeeping can be very rewarding, and it doesn't have to be complicated. Not only does becoming a beekeeper help the pollinator population, but the honey rewards are liquid gold.

Your family and your garden will thank you.

Whether you are young or a little seasoned, like me, bees have been a part of your life since day one. From the food you eat to the flowers you smell, bees have played a role. You may even have a funny story involving a bee—like the time a bee got trapped in the car and everyone was swatting at the air like crazy.

If you know what a bee is and nothing else, but want to learn more about beekeeping, this book is for you. If you are new to beekeeping and want to be a commercial beekeeper, this book isn't written with you in mind specifically, but I still think you'll get some good laughs and maybe even pick up a thing or two that you will find useful.

My desire is for you to have a full, basic understanding of the complete workings of a beehive and bee colony. To make that happen, I'll provide all the information you should need to establish your own successful apiary.

Chapter One

SO, YOU WANT TO KEEP BEES?

Beekeeping is an amazing journey into a whole new world. It's equally beneficial and rewarding for both you and the bees. It can also be as complicated or as easy as you make it. For me, I like the easy path. Don't get me wrong, beekeeping is never totally hands-off. But if you take the more natural approach, you will thoroughly enjoy the benefits beekeeping can bring to your life, without feeling like you are toiling away for your bees.

In this chapter, I will introduce you to some basic bee terms and guide you through everything you will want to consider before making the final decision to become a beekeeper.

What's a **drone?** Who makes the honey? Is keeping bees even legal in your area?

Feeling lost? You won't be after I get you started on the path to beekeeping. I'll help take the mystery out of beekeeping. And I promise I will stick to the basics, so you won't feel stung in the learning process.

Okay, no more bee puns. I promise. Just think about this for a moment: By becoming a beekeeper, you can take an active part in protecting valuable insects. You have the opportunity to make a real difference in the world. Not everyone can make that claim.

THE BEES' KNEES: THE BENEFITS OF BEEKEEPING

Over the years, I've had a lot of different animals and livestock. From my tiny yard to our many acres, bees are hands down the most versatile **livestock** I've ever owned, no matter the size of my land.

Wait, what? Livestock? Yes!

According to the United States federal government, honeybees are classified as livestock (food-producing animals) because the products of beekeeping enter the human food chain, including honey, **propolis**, **pollen**, and **royal jelly**.

colony a collective group of bees that live together

apiary a place where honeybees are kept

drone a male bee; its only job is to mate with the queen

livestock food-producing animals; the honeybee is classified as livestock

propolis also known as bee glue, it's a substance bees make from saliva and botanical sources to protect the hive

pollen a yellow to orange powdery substance produced by plants to facilitate pollination

royal jelly a nutritious secretion produced by the nurse bees and fed to all of the larvae

With other livestock, you may get eggs, meat, milk, or manure. With honeybees, the list of benefits and by-products is a long one.

LIQUID GOLD—AN ABUNDANCE OF HONEY

Oh, the glorious amber sweetness of honey! This liquid gold is the top reason why many people want to become beekeepers. Did you know honey is the only food that never expires? That's right, as long as you keep it covered, it never goes bad. It has a shelf life of F-O-R-E-V-E-R. You never have to worry about using your honey harvest before it spoils. As a matter of fact, the honey you extract this year can be enjoyed by your ancestors hundreds of years from now. How's that for a legacy to leave bee-hind?

How much honey could you leave to your great-great-grandkids? Well, one beehive has the potential to produce up to 60 pounds of honey per season! Don't worry; I promise you'll never be in the situation that you have so much honey you don't know what to do with it. All of your friends and neighbors will happily take any extra honey off your hands.

POLLEN, POLLEN EVERYWHERE

Pollen is actually the sperm cells of a plant—the male part. Plants use pollen to get the male cells from one flower onto another. Whether you love pollen or loathe it, bees need pollen to survive. The bees fly from flower to flower and plant to plant gathering **nectar**, **pollinating** plants, and collecting pollen on their bodies. Some of the microscopic death star (for some of us) particles of pollen are collected by the bees and transferred to other flowers, and some are brought back to the **hive** for food. The tiny particles make their way into the honey, and we consume them.

Beekeepers also collect pollen from the bees to use for health reasons since it's loaded with vitamins, minerals, and carbohydrates. In fact, the German Federal Board of Health lists bee pollen as a medicine. Pollen is just another one of the many by-products of beekeeping that you will come to enjoy.

nectar a sweet liquid produced by flowers that bees collect to make honey

pollination the act of fertilizing plants by transferring pollen from a male plant to a female plant

hive beehives are structures made by people to house bees

HEALTH BENEFITS

There are many health benefits to beekeeping. For starters, you'll live longer. Seriously, science shows that people who have a hobby are 21 percent more likely to live longer. What better hobby is there than beekeeping?

In addition to living longer, you'll have HONEY. Did I mention you'll get honey? I'll share more about this liquid gold in future chapters, but trust me, you'll love the honey.

I'm talking here about raw honey. Raw honey is honey that hasn't been heated, pasteurized, or processed. Unless otherwise advertised, all the honey you buy in the store has been pasteurized. Raw honey is the honey that you will extract from your beehives.

Raw honey has bits of pollen and fine crystals, and, depending on how finely you filtered it, some wax and even the occasional bee part. All the natural yeast and enzymes are intact. This is honey with the maximum health benefits. Pasteurized honey contains fewer antioxidants and nutrients and less bee pollen. Store-bought pasteurized honey may also have some unwanted added ingredients, like sugar.

People consume raw honey, which still contains pollen, to help with seasonal allergies. When they consume local raw honey, they are exposed to small amounts of pollen and can build up an immunity to the pollen of local plants.

FOOD PRODUCTION

After setting up your first beehive and watching your bees, you will start to notice and identify each type of bee and their primary responsibility. The **worker bees** have the job of gathering nectar and pollen for the hive. In doing so, they are pollinating your plants (and all the plants in flight distance from your hive). Plants need this pollination between the male and female plants to produce food and for plant reproduction. I guess you can kind of think of your bees as matchmakers and love connectors. Without bees, our world's food supply would diminish.

Cross-pollination from insects and animals helps at least 30 percent of the world's crops and 90 percent of our wild plants to reproduce. Without bees to spread pollen, many plant species would die off.

BEE-ING A PART OF NATURE

By becoming a beekeeper, you will personally have an active part in protecting an endangered species. How cool is that? That fact alone brings a sense of satisfaction and a kind of joy in watching your bees. Plus, we're still trying to figure out bees' fascinating communication methods. Bees have a language all their own called the bee **waggle dance.** If you watch closely enough, you can observe some of this communication in action. Bees provide an endless source of entertainment for everyone who takes the time to closely observe them.

worker bees female bees; they do all the work in the hive

waggle dance a series of movements that bees use to communicate with each other; the duration and direction of the bee dance can instruct other worker bees where to locate food

THE BUSINESS END OF A BEE: THE CHALLENGES OF BEEKEEPING

Although I really love beekeeping and hope you will too, the truth is, it's not all fun and games. I don't want to burst your vision of dripping with honey and standing in a field of wildflowers, but sometimes things go wrong. I wouldn't be doing you, or your bees, any favors if I didn't bring up the downside of beekeeping.

If you're like me, you may read this portion shaking your head and thinking, "That will never happen to me." Think again. It will happen to you. Maybe not in the same way that it happened to me or other beekeepers, but sooner or later you will face similar challenges.

BEE STINGS

News flash: bees sting. If you decide to become a bee-keeper, chances are you will get stung. If your neighbor decides to become a beekeeper, chances are you will get stung.

This is the most daunting part of beekeeping to me. Why? Because I'm allergic to bee stings. Crazy, right? (I have never met a beekeeper who wasn't a little bit eccentric, myself included.) So why on earth would someone allergic to bees want to be a beekeeper? Well for one, I like to eat, and I need bees if I want to have a productive garden. Two, because I have learned bees don't want to sting you. I will explain bee stings and what to do about them in more depth on page 97, but you should be aware that I've never met a beekeeper who never got stung at least once.

BEEKEEPING TAKES MUSCLE

As I mentioned earlier in this chapter, you can harvest up to 60 pounds of honey per season. That is 60 pounds of honey, plus the weight of the hive and the bees. Not to

Part One

ALL ABOUT BEES AND SETTING UP AN APIARY

mention, the hive may not always be in the best position for deadlifting.

To help with the physical demands of beekeeping, I strongly suggest you connect with other beekeepers in your area or involve a family member or friend to help share the workload.

BEES AREN'T FREE

Bees are free. No, they're not. Before I confuse you too much, let me explain. If you are able to catch a wild **swarm** and you can create a home for them made out of products you already have, and/or a good friend gives you beekeeping supplies, then yes, bees can be free. However, if you aren't fortunate enough to be a part of this unicorn experience, then beekeeping can require a sizeable initial investment. I think it's only fair to share with you the cost of beekeeping, so you can budget accordingly.

Of course, this price will vary considerably depending on what type of hive you buy, what equipment you decide to get, and so on, but our start-up cost for two hives, including bees and without the honey extraction equipment, was around $1,000.

FINDING A BEE BABYSITTER

One of the things I never considered before starting to keep livestock was that fact that I can't just pick up and go any time I feel like it. Unfortunately, this is a problem I run into often. However, the good news is, bees don't keep you housebound the way many other livestock species do. But (yes, there's always a "but") you can't just pack your bags and leave for a three-month trip to the Bahamas and expect to come home to a thriving hive—unless you have someone checking on your bees for you. Finding a babysitter for bees is a little harder than finding one for your toddler. Not impossible, but more challenging. So if

swarm when half of a bee colony leaves with the old queen to form a new colony

you travel a lot and don't have a friend who's willing to wear a **bee suit**, you may want to reconsider beekeeping.

bee suit   a protective suit beekeepers wear to avoid getting stung

BAD THINGS HAPPEN TO GOOD BEES

The worst news of all: Bad things do happen to good bees. Weather, nature, neighbors, disease, poison, and unknown dangers can all destroy your beekeeping dreams. Trust me; I've been there and done that, and it's very disheartening. You can take every safety precaution possible and still lose your bees. It doesn't always happen, but it can happen. Just be prepared to face this reality and not give up. That is the attitude of a true beekeeper.

IS BEEKEEPING LEGAL?

Does this sound like a strange question? I mean, how in the world could bees be illegal? They are the essence of nature, for goodness sake, and without them we would most likely starve. The truth is many cities, especially neighborhoods with Homeowners Associations (HOAs), across the nation, even the world, have rules regarding beekeeping. Some have even declared beekeeping illegal.

Now, you may be shaking your head, ready to dig in your heels if this is true in your area, but I plead with you: Do not move forward with getting bees if they are against the law (or the rules). Instead, I suggest becoming active in your county or city council or your HOA board, and become an advocate for bees; get the laws changed before ordering your bee suit.

HOW TO FIND OUT IF BEEKEEPING IS LEGAL WHERE YOU LIVE

If you live in a community with an HOA, or an apartment building or condo complex, ask to see a copy of the by-laws, rules, or restrictive covenants. Go over these rules and regulations with a fine-tooth comb to see if they address beekeeping. HOA and neighborhood rules can trump city and county laws.

In addition to your community rules and regulations, you will still need to find out the laws in the county and/or city you live in. You can call your county or city building and zoning department, or your county extension office, both of which should be able to help you.

You can also contact your local cooperative extension office or land grant university; they should be able to tell you all of the laws regarding beekeeping in your county.* If you don't have a local extension agency, go online to see if you have a local beekeepers association.

*The University of California website explains how to find your local office: IPM.UCANR.edu/GENERAL/ceofficefinder.html.

 FIND OUT FIRST!

I followed the philosophy of "better to ask for forgiveness later than for permission now" with my chickens and other livestock. Big mistake! The county changed the laws and issued me a citation and an order to remove all my livestock or face jail and $500 a day in fines, plus court costs. It's a long story that involved an attorney, countless sleepless nights, and lots of tears. But in the end, we lost our battle and rehomed all our livestock. Always check your local laws first.

OTHER THINGS TO CONSIDER

You still with me? Good, because I'm almost done with all of the doom and gloom. Like I said before, I wouldn't be doing you, or your bees, any favors if I didn't share the tough side of beekeeping.

Outside of the legalities, stings, diseases, and bad backs, there are some other things you'll need to consider before becoming a beekeeper. I firmly believe anyone getting any livestock, not just bees, should do their due diligence before going out and buying said livestock. Too many animals end up homeless, pockets end up empty, and even marriages end up broken (I kid you not) from lack of research.

WHAT WILL THE NEIGHBORS THINK?

I briefly mentioned bee stings, but this really does need more attention. I've never met a bee that was trained to stay in its own yard. As a matter of fact, bees travel far and wide, which I'll cover in greater detail in chapter 3. My point is, not only do you have to take into consideration that your own family may, and likely will, get stung; you also need to be concerned for your guests and neighbors. With thousands of people, like me, who are allergic to bee stings, it's always good to have signs in the entrances to your property notifying guests and visitors that you have bees. Some stings can be life-threatening. While many people with such allergies carry EpiPens, it's better to be safe than sorry.

 ## INTRODUCING BEES TO YOUR NEIGHBORS

Trust me when I say this—whether you have 1,000 acres or a balcony, neighbors have a direct effect on your quality of life. Good neighbors? Good life. Bad neighbors? Make your life a living you-know-what. Do yourself and your neighbors a favor and be a good neighbor when it comes to your bees.

Here are some tips for introducing your neighbors to bees, and measures you can take to help keep bees out of their way.

» Does your neighbor have a pool? Bees love pools. If your neighbor has a pool, make sure you provide your bees with plenty of water sources, so they are less likely to take a dip in your neighbor's swimming hole.

» Place your hive at the back of your property, at the farthest point from your neighbors.

» Install a fence section or a tall hedge between your property and your neighbors, so your bees will have to fly up and over the obstruction and therefore will be less likely to go visiting.

» Offer plenty of food and water sources on your own property for your bees.

» Invite your neighbors over to share your knowledge about bees, give them honey samples, and answer any questions they have.

» Hand out wildflower seed packets to neighbors for them to plant to help provide food sources for not only your bees but other pollinators as well.

People fear the unknown. If all you know about a bee is that it makes honey and can sting you, of course you're going to be scared. I have learned the best tool to fight fear is education. Share what you know about bees with others, especially your neighbors, so they aren't afraid.

Address concerns your neighbors may have about you keeping bees, and take their feelings into consideration. They may be excited at the thought of a jar of honey from time to time.

PETS AND SMALL CHILDREN

It's easier to keep pets away from beehives than it is small children. In my experience, my pets couldn't care less about the bees and hives; they all just do their own thing. Children are harder to explain things to. If they are older, of course, they can understand the idea of getting stung, but little hands are very curious.

Never leave your child unattended in the yard with the beehives. Keep a medical first aid kit on hand at all times, and do your best to instruct children to stay away from the hives. Don't teach them to fear bees, but rather, to respect their privacy.

Bees are very protective of their **queen** and colony. While they don't want to sting you, they will give their lives to protect one of their own. If they don't feel threatened by you, your children, or your pets, you and your bees can live in perfect harmony. If they do feel threatened, be prepared to either outrun a bee's wings or have your first aid kit handy.

queen an adult female bee that is the only one in a hive capable of laying fertilized eggs

BACKYARD SIZE

Bees are one livestock that has seemed to make the transition from farm life to city life without a hitch. You even see beehives on rooftops in major cities. More important than your yard size to bees is their food source.

GETTING STARTED WITH HONEYBEES

In addition to doing research, like reading this book, getting some basic knowledge, and planning your hive, here are the bare bones of what you need to start your beekeeping adventure.

» **Beehives:** You will need to decide if you want one or more to get you started, what types of hives you want, and where to get them.

» **Supplies:** You will need safety equipment—a bee suit, protective gloves, a hat, and a smoker—and an all-purpose hive tool (more on that in chapter 3). If you plan on extracting honey, you will also need honey extraction equipment, which I will discuss in chapter 8.

» **Food and water:** Before you get the bees, you'll need to set up food and water sources, such as planting flowers and placing water dishes in your yard. You will also need a bee feeder, for when the flowers are not blooming.

» **Bees:** Do you want a pack, a nuc, or a hive that's already established?

You can keep bees on an apartment balcony (provided it's allowed) if there is an ample food source. I will go into bees' nutritional needs in chapter 5; I don't want to get too far ahead of myself. Just note that bees are one kind of livestock that you can raise regardless of the size of your yard or outside living space—which makes them the perfect livestock for almost anyone.

BEES TAKE TIME

Our bees are the least time-consuming livestock we have ever owned. However, that doesn't mean they don't take any time at all. In the beginning, you can plan on investing about 10 hours in setting up your new beehives. Once your hives are established, they shouldn't require more than an hour or two a month, other than honey harvesting season.

During honey harvesting season, depending on how many hives you have, how much honey the hives produce, and how many people are helping you, you can plan on spending an entire afternoon with your hives. We call this day the honey day. If we're lucky, we'll have two honey days a year. The rewards are well worth the time invested.

> **pack**   an actual pack of bees, generally weighing around three pounds; it will contain both worker and drone bees, and should include a caged queen

> **nuc**   an already established frame with wax, brood, pollen, honey, and bees; nucs include drones, workers, and a queen bee

Chapter Two
HONEYBEE 101

Are you ready to learn about the birds and the bees? In this chapter, you will learn about the different types of bees, their role in the colony, and what each type of bee needs to thrive and do to achieve their very important jobs.

With most other livestock, you'll have the males and the females, which are easy to tell apart. Girls have girl parts and boys have boy parts. Bees are more complex, and to the untrained eye they pretty much all look the same. Recognizing the difference between a male and a female bee will take some serious observation. I mean, it's not like you can watch how bees do their business to know if they are male or female. But I will share some tips, so you can quickly identify what sex a bee is by the job that it's performing.

A honeybee colony will be the most well-oiled machine you'll ever have the pleasure of witnessing. Every single bee in a colony has a purpose, knows its purpose, and fulfills it until the very end. You'll never see one bee doing all the work while the other bees hang out by the sunflowers just chilling. Nor will you see a male bee performing a female bee's job, or vice versa. That's just the way it is.

There are three main types of bees that you will have in your colony. They are the queen, the workers, and the drones. Although it may take a little time to train your eyes to properly identify the different bees in your colony, it is extremely important to the success of your hive(s) that you are able to tell them apart.

HER HIGHNESS, THE QUEEN

The queen is the essential, most important bee in your entire colony. Without a queen on the throne, the colony will **collapse.** Having a queenless hive is one of the many ways a beekeeper can lose their hive.

The queen bee is a fully developed female that is responsible for laying all of the eggs in the colony. The queen is the only bee capable of laying fertilized eggs, so

colony collapse disorder an unexplained occurrence when either all or most of the bees in the hive disappear

larva (plural, larvae)   a white, legless, grublike insect that represents the second stage of bee metamorphosis

the entire colony's future is dependent on her survival. No queen? No future bee babies, aka **larvae.**

A beehive with a queen who is producing eggs is referred to as queenright. A beehive without a queen is called queenless. Or a goner, dead, hopeless, empty, bee-lackin' hive—you get my drift. The point I'm trying to make is, the colony needs their lady at the helm. When the queen bee is happy, everyone is happy.

Generally, the queen bee is selected and developed from a larva into an adult by the other bees in the colony. Because of this, the bees in the colony are completely loyal to her and will fiercely protect her at all costs.

QUEEN BEE

IDENTIFYING THE QUEEN BEE

How can you tell who is the queen? There are a couple of distinguishing features.

» **Size**: The queen bee is the largest bee in your colony. This is due in part to her abdomen, which is elongated from carrying the fertilized eggs.

» **Shape**: She will have a distinguishable pointed abdomen. The other bees of the hive will have a more rounded abdomen.

» **Stinger**: Her stinger has less of a bite. The queen bee's stinger is straight and the other bees' stingers are barbed. The sole purpose for the queen's stinger is to fight other queens, not sting humans.

» **Location**: Chances are you won't see the queen outside unless she's getting her groove on. Even a queen likes a little bit of privacy during mating season, and she will not mate inside the hive.

If you're ordering your bees, many times the company or beekeeper will offer to mark the queen bee on the back with a colored dot so you can easily recognize her.

THE WORKERS: THE LADIES OF THE HIVE

The worker bees are the backbone of the beehive. All worker bees are infertile female bees. They are solely responsible for everything that makes the hive tick. Need food? Worker bee. Need water? Worker bee. Need **queen cells**? Worker bee. The worker bees perform every job in a beehive except reproduction.

queen cells cells that are formed when the workers need to replace the queen; they are larger than the other cells

QUEEN CELL

It's quite fascinating to observe the flawlessness and precision that is a bee colony. If bees could rule the world, we wouldn't have as many problems. Each bee has a designated job and performs that job until either the job is complete, or the bee dies trying.

A worker bee will literally work herself to death performing her duty in the hive. If you're walking outside one day and a bee just falls from the sky dead in front of you, chances are she was doing her job until death do her part. Talk about dedication. For the bee, that is life: the survival and continuation of the colony, pure and simple.

JOBS OF THE WORKER BEES

Under the heading of "worker bee," there are many job descriptions, and each worker bee has her own specific job. To understand the complexity of a bee colony it's helpful to understand the worker bees' job duties. Keep in mind, the worker bees are all female.

Some, but not all, of the job duties of the worker bees are:

» **Nurse**: The nurse bees feed and care for the growing larvae.

» **Housekeeper**: The housekeeper bees perform every duty that you would expect a housekeeper to do: They take out the trash, remove dead bugs, clean the comb, and generally keep the hive nice and tidy.

» **Forager**: The foragers gather nectar and pollen.

» **Water girl**: The water girls collect water for the hive.

» **Nectar collector**: These worker bees collect nectar for the hive.

» **Guard**: These are the security guards of the hive and protect it from unwanted guests.

- » **Honeycomb builder**: The construction crew builds the honeycomb, the hexagonal cells of wax inside the hive that hold honey and bee larvae, and repair damaged combs.

- » **Attendant to the queen**: The attendants take care of the queen; they groom her and feed her.

- » **Beautician**: The job of the beautician is to clean off debris and groom the other bees.

- » **House bee**: These worker bees work inside the hive to gather the nectar and pollen from the foragers.

- » **Mortician**: They remove the dead bees from the hive.

In addition to their job duties, the workers are also responsible for the hive's population management. They are the ones whispering in the queen's ear what type of egg to lay. Do they need more workers? "Psst, queen, we need some more sisters. Psst, queen, we need some baby daddies." (They do this by releasing pheromones that communicate this knowledge to the queen.)

 GENDER DETERMINATION

The queen bee can choose whether she wants to have a boy or a girl. Tell me how many humans wouldn't pay a fortune for this superpower?

Once mated, the queen can either lay an unfertilized egg to make a drone, or a fertilized egg to make a worker. The female eggs can develop into either a queen or a worker bee, depending on what they are fed by the nurses during development in the larval stage. This process of choosing male or female is called gender determination, or sex determination. Gender is chosen depending on the needs of the colony.

WORKER BEE

IDENTIFYING THE WORKER BEES

Which bees are the worker girls?

» **Stinger**: Aside from the queen bee, the worker bees are the only other bees in the colony that have stingers. So, if you get stung, it was by a female.

» **Smaller size**: The worker bees are the smallest bees in the hive. The queen is the largest bee, then the drones, and last, the worker bees.

» **Eyes**: Can you see me now? Drones have larger eyes that are closer together and are situated on the top of their head. Worker bees have more of a bridge between their eyes, which are positioned on the sides of their head.

If you see a bee flying around working their little stingers off, chances are it's a female worker bee.

THE DRONES: FLAVOR OF THE MONTH

All drones are males. One might think that a drone's job in the hive would be to protect the queen, guard the hive, act as security, or even sting predators. However, as we already know, the worker bees do all those things. Want to know what the drones' *only* job is in the colony? To mate with the queen. Yes, the drones' only purpose in life is to help create little baby bees.

I feel like I'm still in high school when I talk about this, because I can guarantee if someone had shared this information with my schoolmates, every guy I went to school with would have wanted to grow up to be a drone bee. What a life! That is, until they find out what happens after a night of romance.

Here's a fun fact—well, maybe not for the drone. A drone's male genitalia are attached to their stomach. After fertilization has occurred, the drone's insides are ripped out by the force of ejaculation and it dies shortly after insemination.

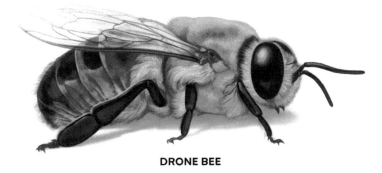

DRONE BEE

IDENTIFYING THE DRONE BEE

In my opinion, the drone bees are the easiest to identify because of their physical attributes.

» **No stinger**: Drones are the only bees in the colony without a stinger. I guess when your only job is to mate with the queen, who has time for fighting?

» **Larger size**: Drones are very similar to the queen bee in size. However, their abdomen is neither elongated nor pointed.

» **Eyes:** As mentioned above, in the worker bees section, the drones' eyes are closer together and larger than the other bees' eyes. This is so they can easily spot the queen during their night on the town.

THE BUZZING HIVE IN NATURE

Bees in the wild will build their nests in hollow trees, crevices in rocks, or most often, in the ground. In fact, of the 20,000-plus bee species around the world, about 70 percent live underground, with only 10 percent of those bees being social and building communal nests.

The behavior and job duties of honeybees remain the same, whether they're wild or in your backyard. You still have the queen producing all the fertilized eggs, the female workers doing all the manual labor, and the drones mating with the queen.

When you decide to become a beekeeper, you are taking nature and placing it in a somewhat controlled environment. While some bees appreciate the security that artificial hives offer, some bees may not and would rather return to the wild. The one thing I can guarantee you about beekeeping: There are no guarantees.

BEES GONE WILD

Bee colonies in search of a new home will intentionally pick an area safe from honey-hijackers. Humans aren't the only honey lovers, so bees are good real estate agents when it comes to location, location, location. Not only do they pick the perfect hiding spot to call home, but they also specifically place their front door facing the south to southeast. Some beekeepers suspect bees do this to maximize daylight flying hours.

Once the **scouts** have found just the right place, the worker bees diligently labor to make their new spot a home. Bees use propolis, which is a sticky residue that

scout worker bees that search for a new source of pollen, nectar, propolis, water, or a new home

they produce, to glue together the materials for their hive or nest. Once constructed, the bees get busy building their **comb** (aka honeycomb). It is within the comb that new bees are raised, and honey is made. The comb is made of lovely beeswax.

Some beekeepers like to take a more natural approach to beekeeping and strive to mimic what bees would do in the wild with their hive placement, such as placing their beehive on the edge of a tree line and facing the entrance south.

> **comb** a group of six-sided cells made of beeswax where bees store their honey and pollen and raise baby bees; the comb is made up of two layers that are attached at the base

 ## HONEY-HIJACKERS

While human consumption of honey dates back thousands of years, there are other species in the wild that love the taste of the liquid gold. These animals threaten the safety of the colony and should be considered a danger to your hive. A few of the honey-hijackers include bears, raccoons, opossums, skunks, and honey badgers. Not to mention all the other insects that are dying to get in, take up residency, and eat rent-free.

BEE COMMUNICATION

Speaking of whispering in the queen's ear, how do bees communicate? Although bees are methodical and meticulous, they are not telepathic, nor are they born with all the knowledge they will ever need to successfully fulfill their job in life. There are things they must learn.

Bees have their own language and way of communicating. It wasn't until 1967 that this bee communication was noticed by scientist Karl von Frisch. He discovered the unique and intricate ways bees communicate through the bee waggle dance, or simply the "bee dance." During a lifetime of study, von Frisch documented how bees can communicate with each other through a series of movements resembling a dance. The direction in which a waggle dance is performed, and its duration, can instruct other worker bees where they can find a food source for the colony.

THE BEE LIFE CYCLE

Understanding the life cycle and dietary needs of bees will better help you raise a healthy colony. Although the bees will do all the work, backyard beekeepers need to do their part in providing food, water, protection, and making sure the hive is healthy.

FROM CONCEPTION

When the queen bee mates with the drones, she stores all the sperm inside of her that she will need for her lifetime. Although it's unknown how many suitors the queen has in her lifetime, studies have shown colonies prefer a queen that has been . . . ahem . . . a little promiscuous.

    With the help of the worker bees, the queen decides whether to lay female or male eggs into the **cells** in the honeycomb (we call these developing bee cells **brood**). These eggs turn into larvae, then **pupae**, and finally, they emerge from their cells as adult bees. The process of development from egg to adult is called complete metamorphosis.

cells   honeycomb consists of hexagonal walls of wax, each enclosing a cell

brood   the developing bees; they might be eggs, larvae, or pupae

pupa (plural, pupae)   the third stage in honeybee metamorphosis, during which it changes (pupates) from a larva to an adult bee

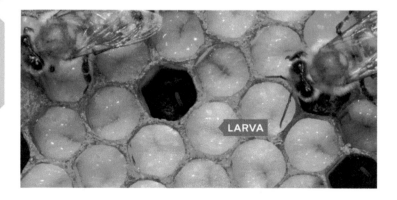

LARVA

frame   racks or frames inside a hive on which the bees make honeycomb

    A beekeeper can tell what bees are developing in the brood **frame** by how the cells look. In the next few pages I will explain the differences in the brood cells, so you can properly identify them and see what the bees are telling you about your hive. Are they making a new queen? Are they producing more male drones? Or is the hive only raising female worker bees?

WORKER BEES: FEMALES RULE

The worker bees will develop from egg to adult in anywhere from 18 to 22 days. During their larval stage they are fed honey and **beebread**, and then their cell is **capped** just before the pupal stage begins.

In a typical hive, you will have 100 female worker bees to every one male drone bee. Consequently, that means when you observe your brood frames you will notice a larger number of female cells than male cells. The cap on the worker's cell is generally flat with very little height to it.

When she has reached the adult stage, the worker bee chews her way out of her cell and is ready to begin working. The average life span of a worker bee is 30 days.

beebread a mixture of collected pollen, bee secretions, and nectar or honey that is a food source for bees, especially for eggs and larvae

capped cells cells that are filled with either brood or honey

 ## WHAT DO DEVELOPING BEES EAT?

Developing bee larvae are fed a combination of honey, royal jelly, and beebread by the nurse bees. Once the larvae are ready to transform into the pupa stage, the brood cells are capped and the pupae do not eat during this stage. At the end of their growing cycle, the bees will emerge as adults from their cells and go right to work.

DRONES: GONE IN A FLASH

The drones' developmental stage ranges from 22 to 24 days. As I mentioned, the female bees greatly outnumber the male bees—not exactly fair dating odds. Because of that, you will notice far fewer drone cells in your brood than worker cells. The capped drone cells will be more dome-shaped, like the nose of a bullet. Their cells are often located toward the bottom of the brood frames, as well.

Drones can't chew their way out like the workers; they must be chewed out by the other bees. Once emerged, the drone is an adult, but will not reach sexual maturity for another nine days. Drones live an average of just a few short weeks. However, they have been known to live a couple of months.

THE BIRTH OF A QUEEN

The worker bees are the ones that determine whether their colony needs a new queen. They relay this message to their current queen by forming a larger cell in the honeycomb, in which she will lay a queen egg. A queen cell looks vastly different than any other cell in the brood. It is shaped more like a peanut and can be found in the middle of the frame, or on the edge.

There are two different types of queen cells, supersedure cells and swarm cells. A supersedure cell is made when the existing queen needs to be replaced for health reasons. The queen produces a **pheromone** that helps guide the workers in their activities. As this pheromone grows weaker due to sickness, disease, or age, the worker bees will know it's time to replace the queen. The workers then develop supersedure cells in the comb, so a new queen can be born. Once the new queen emerges, she searches for her rival, the old queen; they will fight to the death and the victor will take her rightful place at the throne.

Swarm cells are made when the colony has expanded, and they are ready to swarm (half the bees leave the colony to start a new one) to make room in their hive. The workers will develop swarm cells and when the new queen emerges, a portion of the colony will leave with the old queen.

The queen larvae are fed nothing but royal jelly (of course!) and are fully developed in 16 days.

The queen is the longest-living bee of the colony. A queen bee can live up to four years!

pheromones bee pheromones are chemical messengers produced by bees that they use to communicate with one another

QUEEN BEE

Chapter Three

A HOME FOR YOUR BEES

CHOOSING A HIVE

Now that you've learned the basics of bees, the next step in starting your new apiary is choosing the right hive and equipment. If you are brand-new to beekeeping, you may not be aware that there is more than one type of beehive. In this chapter, I will cover all the different types of hives, so you can make the best decision on what type of hive will work for your home and lifestyle.

Some hives require a more hands-on approach, while other hives let you sit back and watch nature do its thing. The hive you choose will depend on what type of beekeeper you want to be. Do you want to harvest and sell bee products, or are you happy just having enough for friends and family? Think about this, because the type of beekeeper you want to be will play a role in the type of hive you should get.

HIVE REQUIREMENTS

If you think of a beehive as you do your own home, it will help you understand some of the basic requirements of a good beehive. Like human homes, bees need a roof, walls, floor, entrance, exit, and ventilation. In addition to structural elements, your hive will need a good location to keep it safe and well-protected. Now that you know the bare necessities, we will get into some of the more detailed descriptions.

RAISE THE ROOF

A roof is a top structure, designed to cover and protect everything below it. I have seen some hives covered with sheets of tin, thatch, wood slats—you name it, beekeepers have tried it. Whatever the material you decide on for your roof, it has to do two things. The first is keep the weather out, the second is to keep bees inside safe from predators. Your hive roof should also be removable for **inspections**.

> **inspection** when you open a beehive to inspect the inside and see what your bees are up to

ROOM TO GROW

Did I mention bees like to reproduce? There's an old saying about rabbits being prolific breeders, but honey, rabbits ain't got nothing on bees. A typical queen will lay between 175,000 and 200,000 eggs each year! That's a lot of bees for one hive. Your bees will need room to grow, either up or out. Some hives will have extra boxes that you can add to the top to make room for expanding colonies. Or you can purchase (or build) additional hives for when your bees outgrow their home.

If your bees don't have room to grow, they will grow out, meaning they will leave their hive in search of a home with bigger accommodations. Typically, half the bees form a swarm and leave with the old queen for better digs. With the help of an experienced beekeeper, you can catch these swarms to place in new hives (thus expanding your apiary), sell them, or let them go in the wild. (Although catching swarms is the least-likely time to get stung by bees, as they are only concerned about protecting the queen and finding a new home, this is definitely not an activity for a beginner. Please let an old hand show you how it's done.)

LOCATION, LOCATION, LOCATION

Location is everything for bees. Again, think of your own home. Are the schools good? How close is the nearest store? What about the water quality? How close are the neighbors? What about the trees, sun, weather? Is the crime rate high? We are still talking about bees, right? Yes. For the many reasons listed below, where you place the hive is of paramount importance, so try to get it right the first time. Just like you, bees don't like to move. Not to mention that moving a hive could increase your chances of getting stung.

» **Food sources**: You will want to plant as many food sources as you can close to the hive, or at least on your property. However, it is impossible to keep bees on your property—no one could ever grow enough food to keep them alive on one lot. They forage up to five miles away to gather enough for them to survive, and even then it's often not enough.

» **Water**: Water is essential. If you don't have any natural water resources for the bees on your land, make sure to place waterers close to the hive and throughout your property.

» **Sun and shade**: Your beehives should not be placed in the direct sun or total shade. A spot that's sunny in the morning and shady in the afternoon is ideal. You also don't want your beehive in direct line of your roof runoff. While natural rainfall is fine, a downpour from your roof is not.

» **Height**: Beehives should be placed at least 16 inches off the ground, to protect your bees from invading insects and other creatures. The stand or support should be very sturdy and weatherproof.

» **Predator-proof**: I mentioned some of the natural predators of beehives on page 23. You will want to avoid any areas that are visited by these honey-hijackers.

» **Children**: Do you have children or close neighbors with children? Then you will want to place your beehive far away from where they play.

» **Neighbors**: Is your neighbor's house, car, or outside living area in the direct path of your bees when they leave or return from their hive? If it is, this may be a problem down the road, even if they love the idea of you having bees. No one is going to like bouncing into bees as they try to get in their car every day.

HOW BEES FIND THEIR WAY HOME

Bees forage for their food a lot farther than you might think. They have been known to fly up to five miles from their home, and even beyond, in search of nectar and water.

Without getting too technical, bees use what's known as optic flow or optic mapping to find their way back home. Basically, optic flow is a mental map insects make to navigate. It is extremely technical, and the scientific advancements that we've achieved over the past decade to figure this out are incredible. The bottom line is that scientists have been able to record and study bees' brains to figure out their internal GPS systems.

Bees use the 5,000 optic lenses in their eyes to record their surroundings and make a mental map of where they are, so they can find their way home. Pesticides have been found to interfere with this process and hinder bees from making it back to the hive. Most pesticides are neurotoxins, and when the bees go to forage, they ingest any pesticides that have been sprayed on their forage. The neurotoxins interfere with their optic flow map, and they can't find their way.

Optic flow is another reason why you don't want to move the hive. If a bee is out foraging when you move its home, it may not find its way back.

WHAT TYPE OF BEEKEEPER DO YOU WANT TO BEE?

The type of beehive you choose will largely depend on the type of beekeeper you want to be. What do I mean by that? Let me go through the different types of beekeepers, and then you can decide which one best describes you.

ENVIRONMENTALIST BEEKEEPER

An environmentalist beekeeper neither wants to produce commercial bees nor sell bee products. They want to keep bees to help the environment. Environmentalist beekeepers are committed to raising bees naturally, without any chemicals or human interference. Their main goal is providing a safe and natural home for the honeybees.

BACKYARD BEEKEEPER

A backyard beekeeper is one who keeps bees as a hobby. Generally, they have one to 39 hives, and may or may not sell some honey or bee products as a side business. This is an official definition; 39 is on the high end, and most backyard beekeepers have fewer than five hives. The backyard beekeeper enjoys the fruits of beekeeping, but not as a full-time income or job.

COMMERCIAL BEEKEEPER

A commercial beekeeper keeps bees to make a living. A commercial beekeeper has 40 hives or more and spends all their time with their bees. Commercial beekeepers often rent their bees out to farmers to pollinate their crops. They also produce and sell bee by-products, such as honey, bee pollen, beeswax, queen bees, nucs, and hives.

BEE BREEDER

Bee breeders are beekeepers who raise bees to sell to commercial beekeepers and hobbyists. Bee breeders also specialize in raising queen bees. Beekeepers have to replace queens (see Requeening in chapter 10) from time to time and need a place to order them from; a bee breeder is that place.

BEE INSPECTOR

An inspector generally works for the state or the extension office and is the first to know about bee problems in your area. This person is considered the local bee expert. A bee inspector should be knowledgeable about all things bee. Eat? Bee. Breathe? Bee. Sleep? Bee. This is the person who is called when a hive has trouble or disease has infected a colony.

WHAT TYPE OF BEEKEEPER ARE YOU?

Now you can decide what type of beekeeper you want to be. If you want to be a backyard beekeeper, or even an environmentalist, this book is written specifically with you in mind. If this is your approach, you'll also have more beehive options than commercial beekeepers. Commercial beekeepers choose hives that will maximize honey production, which limits their options with hive designs. If you want to be a commercial beekeeper, inspector, or breeder, this book will be a good starting point, but you will need to do more extensive research beyond this book.

AN INTRODUCTION TO BEEHIVES

Back in the old days, when people wanted honey they would find a nest in a tree, cut the tree down, and completely destroy the hive to harvest the honey. Thankfully, we've developed methods of beekeeping and honey extraction over the years that don't destroy beehives.

Hives are divided into two categories: removable-frame hives and fixed-comb hives. Removable-frame hives have frames that can be removed for both inspection and honey collection. They have been preferred by most commercial and backyard beekeepers for centuries because their removable frames allow beekeepers full access to their hives for inspection and honey collection. They are also easier to build. Some examples include Langstroth, top bar, Dadant (a top bar variation), and the newest one on the market, the Flow Hive, which is especially designed for easy honey collection.

Fixed-comb hives have frames that are not removable. They tend to be less popular because many beekeepers like to collect honey, plus they are illegal in some states and countries. Some beekeepers still use them because they feel they are a more natural environment for the bees, and they are not as interested in collecting honey. Some examples are Warre, skep (a kind of coil basket), and log hives (yes, a hollow log).

The three most common types of hives are the Langstroth, top bar, and Warre hives, and those are the ones I will discuss here.

THE LANGSTROTH HIVE

The Langstroth hive is the most common beehive among beekeepers, both commercial and hobbyist. They are widely available and easy to find in stores that carry beekeeping supplies. Lorenzo Langstroth patented the Langstroth hive in 1852, but it wasn't manufactured until 25 years after Langstroth's death. The Langstroth hive was influenced by the design of the leaf hive, invented by Francis Huber in the 1700s. The leaf hive had frames that resembled pages in a book, but they weren't removable. Langstroth's hive was the first with removable frames.

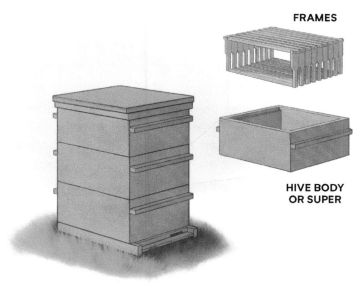

FRAMES

HIVE BODY
OR SUPER

LANGSTROTH HIVE

Not much has changed in the last 150 years from its original design. The Langstroth hive is made up of movable boxes that are stacked in layers. Each box contains removable frames. These boxes are called bodies or hive bodies. The frames are where the bees build their comb. As in all removable-frame hives, the frames can be removed for bee inspection or honey extraction, and then replaced.

In addition to Langstroth's hive invention, he also discovered bee space. Bee space is the ideal amount of cell space bees need to be able to easily move between structures in the hive—and Langstroth discovered that it's $3/8$ inch. At that size (give or take a tiny, tiny bit), the bees leave the space open as a passageway. Any smaller and they regard it as a crack in the hive and seal it up with propolis. Any larger and they fill it in with extra wax. So in a Langstroth hive, there's about $3/8$ inch between each of the frames and between the frames and all the other parts of the hive.

LANGSTROTH HIVES: THE GOOD

Langstroth's invention helped turn beekeeping into a commercial business by giving beekeepers direct access to the honey and wax without destroying the hives, thus allowing future production for repeat harvest. Langstroth hives can be manufactured fairly easily by someone with basic carpentry skills.

Because of their square design and stackable sections, they are easy to transport—another feature commercial beekeepers (who sometimes rent out their bees) like. The square design also allows ease of access, and that has helped beekeepers inspect the hive closely and see when there is a problem, thus reducing disease and infestation in their colonies.

When choosing a Langstroth, you need to decide if you want an eight-frame hive or a ten-frame hive. The difference is the number of frames each section will hold. The eight-frame sections weigh less, since they have fewer frames, making them lighter and easier to lift. It's something to consider if you can't lift a lot of weight.

LANGSTROTH HIVES: THE NOT SO GOOD

Nature isn't square and nature surely doesn't like being manipulated. We often try to take nature and fit it into our lifestyle. Our neat, perfect, square, clean lifestyle. Quoting a scene from *Jurassic Park*, "Life will not be contained . . . Life finds a way."

The predetermined cell size (aka bee space), designed on the **foundations** of the frames forces bees to make specific cell sizes instead of letting the bees determine the size of their cells. Because the frames are oriented horizontally, the bees are also forced to draw their comb sideways instead of down, as they would do in nature.

> **foundations** the wax forms installed in the frames of hives; the bees then build their comb on these foundations

The foundations used in the frames for the Langstroth hives often contain chemicals and therefore can never truly be considered organic. However, you can use frames without foundations if you want to avoid the chemicals that can be found in foundations.

Each section of the **hive body**, when filled with brood or honey, can weigh anywhere from 50 to 100 pounds (depending on how many frames are inside). This heavy weight makes it hard for one person to operate, inspect, and handle the hive.

> **hive body** the main box part of the hive, where the bees live

TOP BAR HIVES

In top bar hives, the frames lift out vertically. They also have one long horizontal hive body instead of stackable sections. They resemble a long box that is either straight like a rectangle (known as the Tanzanian hive) or angled in a V-shape, often with a pitched roof (known as the Kenyan hive). Both types of top bar hives have frames that are accessed from the top of the hive.

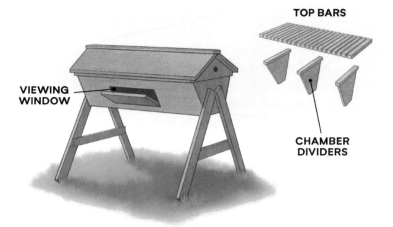

TOP BARS

VIEWING WINDOW

CHAMBER DIVIDERS

TOP BAR HIVE

The top bar hive dates back to the mid-nineteenth century, and is preferred by beekeepers who have back issues or limited strength. It is placed off the ground on stands that make it the perfect height for ease of access.

Unlike Langstroth and Warre hives, top bar hives are just one story high, and the bees' comb hangs down from removable top bars (there's no frame). The bees draw their comb down from the bars placed in the top of the hive, mimicking what they would do in nature.

As the design of top bar hives has evolved, manufacturers have included side windows into the hives, allowing beekeepers to peek at their bees without any disruption to the hive whatsoever. This makes hive inspection less invasive, and the window provides beekeepers with hours of free entertainment.

TOP BAR HIVES: THE GOOD

Top bar hives are preferred by those with limited physical strength. One person can easily handle a single bar full of honey, rather than having to lift an entire Langstroth or

Warre section. They are also at perfect mid-waist height, so they are convenient to access for inspections and honey extraction.

Top bar hives are considerably easier to construct than a Langstroth or a Warre hive. They also have fewer components, making them cheaper to buy. They are very simple in design and operation.

Additionally, the bees draw their comb down from the top bar, making each cell naturally, as they would do in the wild. Their cell size is not predetermined, and the bees are free to create their combs the way they see fit. Since you can remove one bar at a time without removing an entire section of the hive, there is less disruption to the colony.

Beekeepers claim bees in a top bar hive are happier and calmer. Certainly, top bar hives are a more natural approach to beekeeping.

 ## WHAT BEEKEEPERS NEED IN A HIVE

Once you've read about the different types of hives and the different types of beekeepers, you can zero in on what direction you want to take. Your physical condition should play a role in your hive selection decision as well. Are you physically fit and wanting to sell honey? Then you may lean toward the Langstroth hive. Do you want to sell honey and you're physically fit, but prefer a more natural style of beekeeping? If it's legal in your state, then the Warre hive would be a good one for you. Maybe you're not interested in selling honey but just want enough for yourself, you're back isn't what it used to be, and you want to take the natural path. Then the top bar hive has your name all over it.

TOP BAR HIVES: THE NOT SO GOOD

Because of their design, top bar hives do not allow room for expansion; you will need to get more hives if you want to expand.

Some beekeepers claim this design isn't good for cold-weather beekeepers because bees don't like to move horizontally when it's cold. In colder temperatures, bees will stay in the middle of their hive and eat all the honey there instead of going to the combs on the left or the right of the hive to get honey, because it is colder toward the ends of the hive. This means they can't access their food stores in colder weather.

Since the frames for a top bar don't have sides or a bottom, the comb weight isn't supported, the way it could be in a hive with foundations. Because of this, the comb is often damaged during honey extraction and you (and the bees) won't be able to reuse the comb. The bees will have to start over building new comb, which takes time. More time building a comb means a longer time to produce honey.

Since the top bar hives are less popular, it is more difficult to find a mentor or someone with experience to help you. However, they are becoming more popular as the interest in natural beekeeping is on the rise.

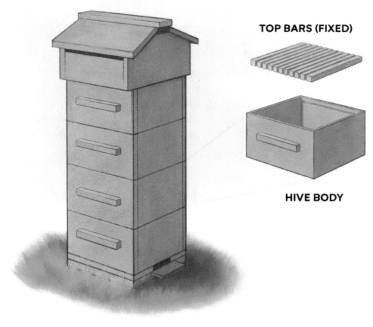

TOP BARS (FIXED)

HIVE BODY

WARRE HIVE

THE WARRE HIVE

A Frenchman named Abbé Émile Warré developed the Warre hive in the early 1900s. Much like the Langstroth hive, the Warre hive is a vertical hive with stackable boxes. However, the boxes on the Warre hive get added at the bottom. Similar to a top bar hive, the Warre hive doesn't use frames or foundations. Rather, the bees draw their comb downward much as they would in nature.

Warré didn't believe in opening the hive for internal inspections, and the top bars are fixed so they can't be removed. The hive was designed to emulate a tree. With the smaller hive design and the frameless hollow boxes, the Warre hive gives the bees a more natural environment in which to build their combs.

One issue all beekeepers have to deal with is moisture in the hive. Moisture and cold can kill bees, so it's very important to protect them (see page 112). The Warre hive is designed to keep cold and moisture out with its unique quilting box.

WARRE HIVE: THE GOOD

Like the top bar hive, the Warre hive is a more natural approach to beekeeping. The foundationless bars allow the bees to create their own cell sizes, bee space, and comb. The **quilting box** design helps absorb moisture and keep bees warmer during the cooler months.

Since the Warre design is based on a hands-off approach, you won't need to perform inspections inside the hive. The smaller design of the Warre hive makes it lighter to handle when you extract honey. Some beekeepers claim that they can collect as much honey with a Warre hive as with the Langstroth hive.

quilting box a box added to the top of the hive that contains a filler that helps absorb moisture from condensation during the colder months, keeping the bees drier and warmer

WARRE HIVE: THE NOT SO GOOD

Fixed-frame hives are not legal in all states or counties (see the box below). Some beekeepers do not like fixed combs because you cannot check on the hive, the queen, or the health of the colony.

Due to the complexity of the design of the hive, construction is more difficult for the novice carpenter. I would suggest buying the Warre hive already assembled.

As the hive boxes fill with honey and brood, they gain a lot of weight. Since you have to lift the hive vertically to add the new box to the bottom, this becomes very hard for one person to maneuver.

As with the top bar hives, the combs in a Warre hive are not supported by a frame, so the comb is more fragile and honey extraction generally means you won't be able to reuse the comb. Which also means it will take the bees longer to produce more honey, because they have to rebuild their combs from scratch.

 THE LEGALITY OF FIXED-COMB HIVES

It is illegal in some states or countries to have fixed comb hives, so please check with your local authority before choosing this type of hive. Fixed comb means you can't remove the comb for inspection. If the comb is not removable, it is not inspectable, and many states feel that is an unwarranted risk of possible contagions.

TO BUY OR DIY?

DIY beehives are not for the beginning carpenter. While they are no Taj Mahal, they aren't as easy as putting Lincoln Logs together, either. My husband and I have been contractors for over 20 years and we still decided to buy our first hives. Why? Simple: time versus money. At the time, we had already ordered our bees and they were on

COMPARING THE TOP THREE HIVES

	LANGSTROTH	TOP BAR	WARRE
Pros	Easy to find supplies Easy to transport Easy inspections and honey extraction Most popular hive, so finding a mentor with Langstroth experience will also be easier	Easy to operate and handle Easy to access honey and bees More natural for the bees Easy to build yourself Happier bees	Natural way of bee-keeping because the bees draw out their comb like they would in the wild Less maintenance and fewer inspections
Cons	Design is not as natural for the bees as the other two hives When filled, the boxes are heavy to lift	Not the best hive for really cold climates May be difficult to find locally	Fixed-comb hives are illegal in some states and countries Hard to build yourself
Cost	$100 to $350	$250 to $500	$150 to $350
Best for	Backyard beekeeper Commercial beekeeper	Backyard beekeeper Environmentalist beekeeper	Backyard beekeeper Environmentalist beekeeper

their way. We were busy with work, and it was too late to even attempt building the hives before the bees came.

BUYING YOUR HIVE

If money is not an issue, or even if it is but you lack carpentry skills or knowledge, then buying your beehive(s) is the best option. There are many benefits to buying a complete hive. One: instant satisfaction. No waiting, no fuss, no muss. Your purchased hive will be brand-spanking-new and ready to welcome your bees home. Ready-made beehives are easy to find for sale locally or online. You can purchase your hive fully assembled, or assemble it yourself in under an hour.

We picked both options (preassembled and assemble it yourself) for our first hives. We bought the unassembled hive first, then after we assembled our first hive, we decided it was worth the extra pennies to order an assembled one for our second hive.

HIVES FOR SALE

Every seasoned beekeeper will have their preferred hive or company they like to use. Although they are all pretty much the same, each hive has minor differences that make them unique to that manufacturer. You can expect to pay anywhere from $100 for a bare-bones hive on up to $1,000 for a hive with all the bells and whistles. However, the average price for a basic starter hive is a few hundred dollars. Some of the top beehive suppliers include Betterbee, Dadant, Kelley, and Mann Lake.

As a bonus to new beekeepers, because backyard beekeeping has become so popular, beekeeping supplies are more widely available. Back in the day, there were just a couple of places where you could buy supplies. Now you can order them online and practically every local feed store sells hives as well. Even our local hardware store carries beekeeping supplies now. Unfortunately, you may be limited to the Langstroth hive if you're buying locally, since most stores don't sell a large variety.

BUYING A USED BEEHIVE

Buying a used beehive is another way to save money. However, if you decide to go this route, I strongly (STRONGLY) suggest you thoroughly sanitize each and every piece. (I'll explain how in Keep Things Clean in chapter 10.) Then when you think you've done a great job, repeat. I'm not being a prude when I say this. Rather, I'm trying to help you protect your investment. Used hives can come with diseases, pests, and chemicals, all of which can be the death of your new colony.

In addition to a thorough cleaning job, you will want to paint your used hive, as I suggested in the box above, to protect the wood from rot.

 ## CHOOSING THE RIGHT WOOD AND PROTECTING IT

Many commercial hives are made out of pine or other soft lumber that is prone to rotting. If you purchase a hive made with these types of woods, you will want to seal your wood with a low-VOC paint and give it plenty of time to dry before adding your bees. (VOCs are volatile organic compounds, which create gases that can be harmful to your bees. Your local DIY store should have paint with low VOCs that is safer to use on your hives.) Make sure to allow several days of drying time so the paint cures and the chemicals have a chance to dissipate.

If you like the wood look and don't want the upkeep of painting your hives every couple of years, choose a hive made of cedar or cypress. A cedar or cypress hive may cost you more money up front, but it will last longer and be more resistant to rot. Cypress and cedar hives are also known for repelling insects naturally, so this is another bonus to this building material.

DIY YOUR HIVE

As I mentioned above, both my hubby and I have many years of construction under our belts and we still found that building our first hives was a bit too intimidating, because we didn't have any real-life examples to go from.

If you have access to an empty beehive and can work your way around a table saw pretty well, you may want to take on this new challenge.

DIY BENEFITS

Making your own beehive will save you a pretty penny, and who doesn't like saving money? You might even want to consider building hives as a little side job. We had a couple of beekeepers in our association who sold home-made hives, and they both used scrap lumber to make them. They charged the same as the commercial manu-facturers, but we didn't have to pay for shipping *and* they were local.

Some other benefits to building your own hives are instant accessibility and long-term growth. If you need a hive now, you can build one, no waiting. A couple of years ago my colony was getting ready to swarm. I didn't have an extra empty hive, nor did I have time to wait for one to be shipped. Thankfully, I had a friend with an empty hive who let me use it. I will explain more about swarms in chapter 10, but if you want to expand your apiary, offering an empty hive for a swarm is the best way to do it.

You may recall that I've mentioned how much bees reproduce. Guess what? All of these baby bees are going to need a place to live eventually. If you only have one box to your Langstroth or Warre hives, you will need to add more as your colony grows. These expansions cost money. However, if you can build them, you can really save on costs and build as many as you need.

MUST-HAVE BEEKEEPING SUPPLIES

The type of hive you decide to get will largely determine what types of tools and accessories you will need. As with the hives, you can be frugal, like me, or you can be all high cotton and buy the top-shelf stuff. Some supplies can be optional (I will explain which ones) and some supplies are mandatory. I mean *mandatory*, as in, you would be crazy

SMOKER

UNCAPPING TOOL

BEE BRUSH

HIVE TOOL

to try to keep bees without it. Not that I'm calling the cops and you'll be arrested for a beekeeping violation. You'll just be one more crazy beekeeper.

HIVE TOOL

Mandatory. You must have a hive tool of some sort to perform hive inspections and to extract honey. The hive tool resembles a pry bar and is great for prying open the hive box, scraping propolis, moving frames, and more.

VEIL

Mandatory. If you've seen those videos of bees crawling all over people's faces while they are singing "Kumbaya" and you think you can be just like them, stop. Just STOP. Bees don't want to sting you, but they will. A bee sting in the eye (yes, it can happen and has to a friend) could leave you blind. A bee sting anywhere is painful, but on the face? Mercy me! Please wear a veil. At all times when working with your hive. ALL TIMES. Nobody likes a bee-stung hero who could have prevented stings by wearing a veil.

VEIL

BEE SUIT

SMOKER

GLOVES

HIVE TOOL

It's also a good idea to have a second one in case you need a helper.

BEE SUIT

Mandatory. Or optional. This is one of those should-have items, but it's not 100 percent necessary. I highly advise you buy one. But if you don't, you must wear full-body protective clothing. It's also a good idea to have a second one in case you need a helper.

GLOVES

Optional. Bee gloves come in heavy cloth, thick rubber, cowskin, or goatskin. You may decide not to purchase gloves; I wear them every time. I have worn the cloth and was stung through them, but a stinger has yet to get through my goatskin gloves. (Knock on wood.) A frugal suggestion is those yellow dish gloves. Get the thickest ones you can find. They won't protect you like bee gloves, but they are better than nothing in a pinch.

SMOKER

Mandatory. Or optional. A smoker is a small can with a bellows and a spout. You make a little fire in it, and use the bellows to make smoke (read more in chapter 6). It's used when inspecting the hive or extracting honey. The smoke makes bees less active and reduces the chance of aggressive bee encounters. I do recommend a smoker for the beginner beekeeper who is not familiar with handling bees. However, I do have many beekeeping friends who do not use a smoker with their hives.

BEE BRUSH

Optional. If you've ever seen an ice scraper for windshields that has a long brush on the end, that is what a bee brush looks like. It's great for gently brushing the bees off a frame so you can inspect it or extract the honey. I think purchasing a bee brush is optional but I highly suggest you purchase one, especially because they can be bought for under $10.

REFRACTOMETER

Optional. This is a tool used to measure the moisture level in a hive or in the honey. Excessive moisture in a beehive can kill your bees. Some beekeepers use a refractometer to measure the level of moisture in their hives. Another way to use it is to measure the moisture content in your honey (see page 119).

EXTRACTOR

Optional. An extractor is a machine used to extract honey by spinning the frames. If you want to sell honey, this is mandatory. However, you can still extract honey without a honey extractor. In fact, you can even make your own if you watch enough online videos. If you get friendly enough with another beekeeper who has one, they may let you borrow theirs. Since extractors are only used a couple of times a year, beekeepers are generally open to sharing. Our beekeepers association had one for its members to use for free. Another thought is to go in on the purchase with another beekeeper to reduce the expense.

BEE FEEDER

Mandatory. There will be a time, or times, where your bees will need to be fed. During these times you will need some type of bee feeder. Some feeders are made to go inside the hive and others go outside. Outside feeders will probably be inaccessible to bees during bad weather and when it's cold, so you probably want one of each. They aren't expensive, so this is a fairly painless purchase.

WATER DISH

Mandatory. Bees can't live without water. You will need to offer shallow water dishes or birdbaths for your bees. The bees need shallow dishes so they don't drown. It also helps to place something in them that bees can rest on, such as clean rocks. I used a birdbath and placed marbles in it—until my bees discovered they liked our natural pool better.

FRAME GRIP

Optional. A frame grip is a great big pair of tongs that helps you remove the frames from your hive box. You use the frame grip with any hive that has removable frames. I've never had one, but I know those who have. While they are a nice addition, they are not mandatory.

Chapter Four

PLANNING YOUR COLONY

Bees are bees, right? If you are a new-bee bee-keeper, you may not even be aware that there are different types of bees. There are almost 20,000 different types of bees, to be exact, and they are found on every continent except Antarctica (where it's too cold and there's nothing for them to eat). Thankfully, you don't have to choose your honeybees from a catalog with 20,000 options.

In the United States there are six main honeybee races. (All honeybees are the same species, but they come in different subspecies, called races or stocks.) They are Italian, Carniolan, Buckfest, Russian, German, and Caucasian. In this chapter, I will go over the three most popular honeybee races: Italian, Carniolan, and Russian. I will also discuss what types of plants you should add to your landscape to make all your bees happy.

WHICH BEE IS BEST FOR YOU?

Are you a Ford or a Chevy kind of person? Or are you more like "as long as it drives, I don't care what it is"? If you've ever been in the middle (or at the end) of this question, you know people can be loyal to their favorite brand. The same goes for honeybee races. Gather a room full of beekeepers and ask them what their favorite bee race is. You'll have a multitude of opinions—all of them right.

When I started beekeeping, we ordered bees through our beekeepers association and they only had Italian bees, so that is what I got. Depending on your circumstances, you may not have a choice in the type of bee you get. However, you can generally find all three races available for sale online.

If you go back to the section What Type of Beekeeper Do You Want to Bee? (page 32), not only will that help you decide what type of hive to get but it will also help you in choosing what type of honeybee you want to buy, based on their characteristics. Similar to the different beehives, each bee race has characteristics that will appeal to particular types of beekeepers.

For instance, if you want a gentle bee that's not prone to swarming, then the Italian honeybee would be best for you. If you want really resilient bees that handle the cold well, then the Russian honeybee would be good for you. If you live in an area with a lot of wildflowers and trees, then the Carniolan honeybee might be the one for you. Much of this decision will be based on availability in your area, too.

ITALIAN HONEYBEE

The Italian honeybee (*Apis mellifera ligustica*) is the most popular among beekeepers. They originated from Italy but have adapted to survive in climates all over the world. They are known for their gentle nature and they aren't prone to swarming—although they do swarm. They are strong honey producers and great beginner honeybees.

The downside is although they've adapted to many climates, they don't thrive in high humidity or places with extreme cold. They make horrible neighbors to other

COMMON BEE RACE ATTRIBUTES

	ITALIAN HONEYBEE	CARNIOLAN HONEYBEE	RUSSIAN HONEYBEE
Docile	Great	Great	Good
Isn't prone to swarm	Great	Good	Okay
Good honey producers	Great	Good/Great	Good
Overwinters well	Good	Okay	Great
Disease-resistant	Okay	Okay	Great

beehives, because they like to rob honey if food is in short supply. In addition, Italian honeybees often get lost and have a hard time finding their way back home.

CARNIOLAN HONEYBEE

The Carniolan honeybee (*Apis mellifera carnica*) is the second most popular honeybee among beekeepers. They are a great honeybee for producing worker bees when there is a heavy **nectar flow**, and slowing bee production when there is less food available. Carniolan bees are extremely docile but are great defenders against pests.

nectar flow the time of the year when flowers produce nectar

Their honey production is less than that of the Italian bee, and they are more prone to swarm. They aren't fans of hot summers. Then again, who is?

RUSSIAN HONEYBEE

Russian honeybees (*Apis mellifera*) were introduced into the United States in the late 1990s to increase the declining bee population. They were introduced because they are resistant to parasitic mites that have plagued honeybees. They are not an aggressive bee and they overwinter well.

The downside is they do tend to cost more than other bee races and are prone to swarming. If they do swarm, you could lose your investment.

BUYING YOUR BEES

So the big question is where and how does one buy bees? It would be easy if your local grocery store carried them. Unfortunately, it's not quite that easy. When buying bees, you have several options. But first, you need to decide how many bees you want to buy and how you want them to come. Then there are always FREE bees too, but I'll get to that in a bit.

There are three different ways you can order bees from bee breeders or companies. They are packs, nucs, and full hives (also known as established colonies). My husband and I bought packs, and nucs, and we obtained feral swarm bees. I wanted to buy full hives but could never find one, due to the limited availability where I live.

You will only need to buy one pack or nuc for each beehive you own. If you buy a full hive, that is all you will need until the hive grows and you have to add more living space for them.

PACKS

Packs are generally three pounds of packaged bees, and will include, drones, workers, and a mated queen bee—meaning some drones have already visited and she is pregnant, so to speak. The queen that comes with packs is generally not the colony's queen, and it will take some time for them to get to know and accept her. Packs are priced from around $50 to $80 in most areas and contain around 10,000 bees. (The basic rule is 3,000 to 4,000 bees per pound.)

Packs are a great bang for the buck, but will take longer to get established since they are just the bees without any comb or brood. However, many new beekeepers enjoy seeing a colony develop from the ground up, so to speak.

NUCS

Nucs are already established drawn frames—frames with wax, brood, pollen, honey, and bees. Nucs include drones, workers, and a queen bee. The queen sold with nucs is often the colony's queen, so they have already accepted her and are in the stage of making brood (bee babies) and honey. Nucs can run around $100 on up and also contain around 10,000 bees, depending on the seller. The average nuc contains five frames. Nucs are worth the investment because they are established and ready to start producing as soon as you get them.

Nucs are only used in Langstroth hives, so if you are going with a top bar or a Warre hive, you will need to purchase packs.

ESTABLISHED COLONY

Established colonies are harder to find and a bit more expensive but remain a fantastic option to get you going right away. They consist of one full hive body, which is 8 to 10 established drawn frames with wax, brood, pollen, honey, and bees. They come with the drones, workers, and a queen bee that is the queen of that colony. Established colonies are the easiest way for new people to become beekeepers.

Some of the downsides to purchasing a full hive are the cost (about three times more expensive than a nuc), availability (the lack of it), and shipping. You should do an inspection to make sure it's a healthy colony. You will want a second opinion from a seasoned beekeeper to help you with the inspection of an established colony, to make sure you are getting healthy bees.

Any reputable beekeeper or breeder will offer some type of health guarantee with purchase—but not everyone is that reputable. Check with the company or business you are ordering your bees from and ask if they have a health guarantee and what to do if your bees arrive sick or unhealthy.

FREE-BEES

Is there such a thing as free anymore? That answer would be yes, if you don't charge for your time. There are a handful of ways you can get bees for free. However, I tend not to recommend these ideas to the very beginning beekeeper, because most of my suggestions involve having some bee experience.

SWARM CAPTURE

During certain times of the year, honeybees swarm and split from their colony to form a new colony. These swarms are looking for a new place to call home and multiply. Beekeepers love catching swarms because they are free-bees. Spring is when bees will swarm, so if you want to get bees this way, spring is the time.

If you are interested is catching a swarm, I suggest doing so only under the supervision of an experienced beekeeper. A brand-new beekeeper should never attempt to capture a swarm alone, especially without any prior experience. An experienced beekeeper will help you and the bees stay safe during the process. They will also be able to make sure you have captured the queen when you relocate the swarm inside your hive. Without the queen, you won't have bees for long. (See A Royal Coup D'état in chapter 10.)

OTHER BEEKEEPERS

I know this is hard to believe, but sometimes beekeepers are looking to give away bees. It may be that they only wanted a small apiary and their hives are growing too big, or maybe they wanted to get out of beekeeping. You need to be in the right place at the right time—and how often does that happen?

BEE REMOVAL

Bees don't always live in neat little beehives. Sometimes they think an attic or a wall in a house would make a good home. Pest control companies and local extension offices get these calls all the time, and they need a place to relocate the bees. If this is an option you would like to explore, give them a call to see what you can work out.

WHEN TO BUY BEES

Bees tend to slow down and take the winter off, but come spring they are ready to get out and get busy. There is a golden time—right between the winter slumber and the spring flowers—that beekeepers think is the perfect time to buy new bees.

During the winter, bees become very inactive. There is very little to forage on and they huddle inside their homes to keep warm. In the winter, they eat much of their stored food, and by spring are near starving. They rely on your spring flowers to jump-start their new year of honey production. You will want to order your new bees in the spring when there is plenty for them to forage on and before it gets too hot.

If you missed spring and still want bees, don't worry. Spring isn't the only time to get bees, just the best suggested time. You can buy a full hive any time of the year.

A HONEYBEE GARDEN

Part of your responsibility as a beekeeper is providing food for your bees, by planting flowers they can forage on, and offering supplemental feeding when there is no forage

available. Bee gardens are aesthetically pleasing and provide a dual purpose. Bees and gardens are the perfect balance of life. The bees need the flowers to live and the plants need the bees for pollination.

When I plant for my bees, I like to take a couple of things into consideration. First, is the plant a perennial or an annual? I like to plant things that will return year after year; less work for me and money saved. Second, will it serve more than one purpose? Meaning can I cook with it, use it for medicine, or some other use? I love dual-purpose things, including plants and livestock.

Another thing to take into consideration when picking plants for your pollinator garden is to make sure you have plants that bloom at different times of the year. You'll want some plants that bloom in early spring, some in the summer, some in the fall, and some in the winter if you live in a planting zone that doesn't freeze in winter.

 ## FLAVORING YOUR HONEY

Have you ever eaten a lot of garlic or onions and then noticed that your breath or even your skin smells like it? The same goes for bees; whatever bees forage on, that is what their honey will taste like. Beekeepers use this knowledge to their advantage and come up with an array of different flavors of honey.

I have a friend who owns a blueberry farm and got into beekeeping to help her berry production. Her blueberry-flavored honey came into such high demand that her berry farm took a back seat, and now she is a full-time beekeeper selling her blueberry-flavored honey.

This concept also works in reverse. In the fall we have a lot of goldenrod, and this is generally the last crop the bees will forage on before winter. Some people love the taste of goldenrod honey, others detest it. For this reason, you will see beekeepers advertising fall or spring honey, so experienced honey customers will know what the honey tastes like.

PICK YOUR PLANTS CAREFULLY

When planting for pollinators, it's important to buy plants that haven't been treated with toxic chemicals for bees. If the plant advertises "disease resistant" or treated with *anything*, don't buy it for your garden. Chemicals have a toxic effect on pollinators and will weaken, if not destroy, your bee colony.

Instead, stick to organically grown plants. Better yet, start your own from seed. Remember this even after you put your plants in the ground: Don't spray them or treat them with chemicals. These are your bees' source of food and they need their food to be chemical-free.

HERBS, FLOWERS, AND GROUND COVER

When starting any garden, it's important to know your plant hardiness zone. You can look online (try the USDA: PlantHardiness.ARS.USDA.gov/PHZMWeb) and enter your zip code, or contact your local county extension office to find your plant hardiness zone. What grows well in my zone of 6b may not do well in your plant hardiness zone. A local nursery should be able to guide you as well.

The plants listed below are just a few suggestions that your bees will love. Before you buy them, make sure they will grow well in your area.

GROUND COVER FOR BEES

Calico aster	Lanceleaf coreopsis
Clover	Lanceleaf self-heal
Dandelions	Mint

HERBS FOR BEES

Bee balm	Lavender
Borage	Rosemary
Echinacea	Sage
Hyssop	Thyme

FLOWERS FOR BEES

Black-eyed Susan	Honeysuckle
Goldenrod	Sunflowers

Part Two

THE FIRST YEAR OF YOUR NEW COLONY

Are you ready to get your bees and get started? Just picture your hive in your backyard: flowers are blooming, your bees have been ordered, and you're anxiously awaiting their arrival. Now what?

The "now what" is what I will cover in this part of the book. You will learn how to transport your bees home, introduce them to their new hive, and make sure they don't fly away. In addition to the homecoming, I will walk you step by step through your first hive inspections and subsequent inspections. Let's not forget the possibility of getting stung by your new livestock. OUCH! I've got tips for that too. You can greatly reduce your chances of getting stung if you suit up properly—which is also covered in this part.

Undoubtedly, you're excited to start swimming in honey, but we still have some things to cover to make you a successful beekeeper. Beekeeping is a hefty little investment, of both time and money, and I want to make sure you have all the knowledge you need to do it right.

Chapter Five

BRINGING HOME THE BEES

Whether you've ordered packs, nucs, or a complete hive, you will learn about the steps of bringing home your bees and getting acquainted with their new surroundings. When you purchase bees directly from a bee breeder, local business, or beekeepers association, they will inform you when and where to pick up your bees. If you are ordering your bees online, they will be delivered to your post office; bees do not get delivered to your front door. After you have ordered your bees online and have received an estimated delivery date, inform your post office that you are expecting them. When they arrive at the post office, you will need to pick them up first thing in the morning (or as soon as they arrive). Do not let them sit at the post office all day waiting for you to get home from work. Nobody will be happy—not the postal workers, and not the bees. Plan ahead and make sure you have this day available for pickup.

BEE PREPARED

Wherever you get your bees from, they should keep you informed about the delivery date. When your bees are close to being delivered or picked up, be sure you have:

» Checked all the state, local, and neighborhood laws and verified that it's legal to keep bees where you live

» The hive (If you purchased a used beehive, the hive has been thoroughly cleaned, sanitized, and dried. If you bought a pinewood hive, the hive has been painted, or sealed, with plenty of time to dry.)

» Completely assembled the hive and placed it in its permanent location

» All your safety gear, such as your veil, suit, and gloves

» Hive tool and bee brush

» Smoker and fuel for your smoker

» Bee feeder in place

» All water dishes in place

» Informed your neighbors that you are getting bees

» Registered your apiary with your county or state (see Registering Your Beehive on page 67)

HOW TO TRANSPORT BEES

If you ordered your bees, chances are you will have to transport them home. This means you will be bringing thousands of bees from their location to your location. Whether by plane, train, or automobile, your bees will need to hitch a ride home. Although bees do come packaged, there are usually a few hitchhikers that didn't make it into the box with the rest of their buddies.

I have a funny story about this, but first, the vehicles. The best vehicle to transport your bees is something with an open bed, like a pickup truck. You can place your bees in the trunk of a car, but only if the temperature is not too hot. If a car is left in the direct sun, the internal temperature can reach more than 104°F on a day that's only 70°F outside. If it's too hot for you to sit comfortably (temperature wise; don't actually sit in your trunk, that would be silly—and unsafe), then it's probably too hot for your bees. SUVs or vans work, too, and the bees will then be in the part of the vehicle that's climate-controlled. But you will still need to take some extra measures for transport, which I will explain in the next section.

TRAVEL TIPS

As with any livestock, transport to a new home can be stressful. Your goal is to minimize that stress as much as possible for your bees.

REGISTERING YOUR BEEHIVE

Some states now require you to register your apiary. Where we live, we have a law called the Apiary Act. We legally have to register our apiary with the state Department of Agriculture, and we have to reregister our apiary every three years.

Aside from it being the law in some states, there are benefits to registering your bees. The State of Tennessee's apiary registration website explains what some of the benefits are—and the consequences if you don't register.

There are a number of benefits to registering your apiary:

» E-mail notification of disease outbreaks and updates from the State Apiarist.

» E-mail and postal notification of aerial spraying of pesticides in your area when we are notified of the spray-ing projects.

» Free inspection of your colonies if you are selling them, moving them, or you feel you may have a bee health problem.

» Registering your bees helps to protect your bees and your neighbor's bees in the case of an American Foulbrood (AFB) outbreak or other regulatory pests.

» If your colonies have to be destroyed due to American Foulbrood or other regulated pest or disease, you will be compensated if they are registered. There is no indemnity paid for the loss of unregistered bee colonies.

What can happen if you do not register your bees or your apiary?

» Failure to register your bees or comply with the provisions of "The Apiary Act of 1995" may result in the confiscation of your bees, beekeeping equipment, and a $500.00 fine.

» If your colonies have to be destroyed due to American Foulbrood or other regulated pest or disease, you will not be compensated if they are not registered.

BRINGING HOME MY BEES: A CAUTIONARY TALE

Being a first-time beekeeper, I had no idea what to expect when I picked up my bees at our beekeepers association meeting place at 1 a.m. We didn't have a pickup truck back then, nor did I know enough to ask to borrow one. So here I am, driving down the highway at 1 a.m. with a trunk full of thousands of bees. Little did I know that our old car had gaps between the back seat and the trunk.

Before you know it, my teen was shouting and the car was filling up with bees (the hitchhikers), and they were flying all around. As I have mentioned before, I'm allergic to bee stings. Well, sure enough, one stung me. It landed on my steering wheel and I couldn't see it, so I accidentally squished it with my hand. I immediately pulled off to the side and jumped out of my car and began running around it trying to escape the bees. I can only imagine what passersby thought. They were probably tempted to call the police. Learn from my mistakes and follow the travel tips below.

» Skip the errands on bee pickup day. In other words, go straight to where they are and head straight home.

» Use a pickup truck, if you can, and put them in the flatbed (not the cab).

» If you're using a car, SUV, or van for pickup, bring a light sheet or pillowcase to cover the bees, so they don't end up flying around the car. A clean mesh bag, like what you would put dirty clothes in, is even better—as long as the mesh openings are smaller than honeybees.

» Drive like you're teaching driver's ed: No sharp turns or sudden stops.

» Provide airflow, especially if you will be traveling a long distance with your bees in the trunk. Rig something up so you can safely keep the trunk open a tiny bit.

WELCOME HOME

You have your bees in tow and you've made it home. Congratulations! You're now a beekeeper! You are officially part of a secret society that is dedicated to preserving this quintessential part of the web of life. Furthermore, you get honey.

The next step is to transfer your bees from their transport container to the hive you set up for them. You can't just dump them in. Bees are sensitive little beeings!

THE BEST TIME TO MAKE THE TRANSFER

If you bought a full hive, you can skip this part. However, you may need this knowledge in the future, so keep reading.

Bees are most active during the day. They're out foraging, collecting pollen, nectar, and water. Your bees' first inclination, if you transfer them during the day, is to go do what bees do best. If your bees take off to go work without really knowing where home is, chances are they won't come back. Therefore, it is best to rehome your bees at dusk or dark.

If you picked up your bees in the morning or afternoon, leave them in their transport box and place it in a shady spot, safe from winds and drafts. You can lightly mist the box with **bee food** that you have poured into a clean, never-used-before spray bottle. This will keep them fed and hydrated until it's later in the day.

bee food a sugar water supplement used to feed your bees when nectar is not available

PREPPING YOUR HIVE

If you've bought a Langstroth, or another hive with frames, you'll want to remove two frames to make room for the bees. For example, if you're using a 10-frame hive body, remove two of the frames from the center and space out the remaining frames. If you've bought a hive with only top bars and no frames, you will just need to remove some of the bars to allow space to pour your bees inside.

USING AN ENTRANCE REDUCER

An entrance reducer is a small block of wood that reduces the size of the entrance to the beehive. To my knowledge, only Langstroth hives use entrance reducers. However, if you see the benefit of using a reducer for a hive other than a Langstroth, you can make one by simply placing some twigs in front of the entrance of the beehive (make sure not to block it entirely).

A reducer is used in a couple of different scenarios.

» **Installing new bees:** A new colony will not be large enough to defend itself against predators and pests. Reducing the size of the entrance creates a smaller opening for the bees. After a couple of weeks, you can remove the reducer during your second hive inspection, if you notice your bee population is growing.

» **When it gets cold:** Many beekeepers who use Langstroth hives use the reducer during the colder months to reduce drafts.

» **When fumigating with various chemicals to disinfect or treat disease or pest infestations:** If you're fumigating the hive with essential oils or other pest treatments, a reducer is used to keep the fumes in the hive.

However, there are problems with reducers. Recent research suggests that using entrance reducers during the winter months creates excess moisture, which kills bees. Instead of using reducers, some beekeepers suggest placing small twigs across part of the opening but not so many that the opening is blocked.

Some beekeepers forget to remove their reducers during the heavy honey flow, thus limiting how many bees can use their door at a time. In addition to crowding the exit, failing to remove the reducer during warmer months will cause your hive to overheat.

Use caution when using an entrance reducer, and remove it before the warmer months so your hive doesn't overheat.

SUITING UP

Safety first. I can't stress this enough. You can work on becoming the bee whisperer later, but for right now, I want you to look like you're going into a hazmat operation. This means gloves, bee suit, veil, and boots. Getting dressed is a one-person job, but it's always good to have a second pair of eyes to make sure you did it right.

You want to cover every inch of exposed skin and block every opening under your clothing. If there is the tiniest of cracks in your defenses, your bees will find it. I tuck my pants into my socks and I pull up my shirt collar around my neck—before putting on my bee suit. After I'm ready to go in, I have my husband give me a once-over to see if I've missed any valuable real estate that the bees will take advantage of.

START SMOKING

I will say it one more time, so you know how serious I am: Before handling your bees, make sure you have all of your safety equipment on correctly. In addition, you'll want to fuel up and light your smoker, and get your bee brush ready.

You will need to use your smoker to install your new bees—unless, of course, you decide to go the all-natural route. I will go over step-by-step instructions as well as the ins and outs of the bee smoker on page 88. Make sure to check on your smoker from time to time while you are busy with the bees.

honey flow   the time when bees produce a large amount of honey

RELEASE THE BEES!

Your bees should come with instructions on how to unplug or release the queen and transfer your bees to the hive. In the case that didn't happen, or the instructions are not clear, let's go over the basics of installation. Before you start, your beehive should be prepped and ready to accept your new bees, with the cover off and some of the frames removed to make room for the new arrivals. Of course, you'll place the queen inside the hive before you install the rest of the bees.

PRESENTING HER MAJESTY, THE QUEEN

CAGED QUEEN

caged queen   a queen that is not the colony's queen, and can be bought separately; she comes in a cage and must be freed by the worker bees

Your pack of bees should be in a temporary box, with a **caged queen**. The queen's cage is generally made of wood on two sides and wire on the other two sides. The queen bee is trapped in her cage by a candy plug—a sugary plug sealing the bottom of the queen's cage. Because it is sweet, it entices the bees to eat it and release the queen.

The first thing to do is to remove the queen's cage from the package of bees or the nuc—carefully. Gently use your bee brush to remove any adoring bees on her cage, so you can inspect the queen. Make sure she is moving around and appears healthy. If she isn't moving, contact your bee supplier immediately.

The queen that comes with packs will most likely not be the colony's naturally chosen queen, so they will need time to get to know her. Her Majesty will not like being trapped in a small cage and will release a pheromone. The worker bees will respond to her pheromones and release her by eating through the sugar plug. That process is how the colony gets to know their new queen. (See What If the

Queen Hasn't Been Released? on page 83 for . . . what do to if the queen hasn't been released.)

Your instructions should tell you to remove a frame in the middle of the hive and place the queen where the frame used to be. Secure the queen cage to an adjoining frame with a paper clip, or you can use a rubber band to hold the queen cage tightly against the frame. If you have a frameless hive, you can attach the queen cage to the top bar or to the side wall with a paper clip.

After you've finished installing the queen, you can then install your bees in the hive.

INSTALLING A PACK OF BEES

Have your hive tool, bee brush, and smoker ready. You may need your hive tool or a pry bar to help you access your bees. Generally, packs of bees will have a piece of wood or cardboard stapled over the opening. Right below this cover will be their travel food, and after the food are your bees! This is when you will want to use your smoker.

» Lightly smoke your bees *before* you remove anything from your bee box. This will help calm them down before you add them to your hive.

» Use your hive tool to remove the covering and travel food to expose your bees.

» Gently tip your bee box upside down and lightly shake the bees out into the hive.

» Several bees will remain in the box; that's okay.

» Place the empty-ish box next to your hive and leave the box there for several days or a week.

» Use your bee brush to gently brush the bees into the hive so you can place your cover on top.

Congratulations! You've successfully relocated your bees to their new home. Now just breathe and relax.

NUC

INSTALLING A NUC

Nucs can only be installed into a Langstroth hive, unless you built your own hive that the Langstroth frames will fit into. Your nuc is a miniature version of an established colony. You should receive two to five frames complete with brood and honey, along with drones, workers, and a queen bee. Your nuc should also come with a queen that belongs to the colony. However, if your queen is in her own cage, you will need to follow the instructions on page 72 for releasing the queen (and remember, Her Majesty goes into the hive first).

» Start by taking off the hive cover and removing two more frames from the center of your hive box than what your nuc has. For instance, if your nuc came with five frames, remove seven frames from the center of your hive box. Bees like to work from the center out, so always place the bees on their frames in the center of your hive when installing.

🐝 LEMONGRASS OIL

Lemongrass essential oil has gained popularity among beekeepers looking for natural ways to keep bees. Beekeepers use lemongrass essential oil in their new hives, so bees will be drawn into their new home. Here's how it works.

Bees use their Nasanov gland, located in the abdomen, to release pheromones to guide other bees to a new location or to a food source. In other words, bees use their Nasanov gland to mark their territory, much as dogs do with their urine. Lemongrass contains two of the natural fatty acids that are found in the pheromones bees use for marking their new home.

When using lemongrass essential oil in your beehive, make sure just to put a couple of drops, no more than that, in the inside corners before you release the bees. You want to attract the bees, not fumigate them.

》 Lightly smoke your bees to help them calm down. Next, carefully remove the frames from your nuc and place them in your hive in the same order they were placed in the nuc.

》 Gently tap any remaining bees from the nuc packaging into your hive and place the box next to your hive box. Any remaining stragglers will find their way home soon.

》 Take your bee brush and carefully brush the bees so they are all (or mostly) inside the hive, and replace the top of your hive box. Make sure your bees have food and water.

INSTALLING AN ESTABLISHED COLONY

If you've bought an established colony, your job of installation is pretty easy. You should receive installation instructions with the purchase of your hive. If they differ from what I'm telling you, please follow the supplier's instructions.

Figure out where you want to place your hive before it arrives. When it comes, just set it up on its foundation (something that will keep it a little bit raised off the ground). This spot should be its permanent location.

REORIENTING A HIVE

There's a saying in my beekeepers class: "You can move a hive three feet or you'll have to move it more than three miles." It means you can move your hive just a little bit and the bees will be fine; they can fly away from the hive and still find their way home with no problem. Or you will have to move your hive more than three miles so they will reorient. Others say two feet or two miles. I'm not sure how much

truth there is to this, but I've never put it to the test.

It is true that when you relocate bees to a new area, they will need to reorient themselves (see How Bees Find Their Way Home on page 32). If you don't help them reorient, they will fly off to forage and not be able to find their way back. You can help your bees reorient by keeping them confined to the hive in their new location for 72 hours—whether you've set up a new hive

or just moved an established one. This will help reset their internal GPS systems.

You can confine your bees to their new location by blocking the entrance to their hive with twigs, grass, or a small piece of wood or mesh wire. Of course, they must have food and water, so you will have to supply that inside the hive. After the 72 hours are up, remove the blockage from their entrance and let them free to begin their orientation flights.

orientation flights when new bees take their first flights out of their hive to get to know their surroundings

After your established hive has been placed off the ground in its perfect spot in your yard, keep the bees confined in their hive for 72 hours. (Talk to the person or company that you purchased your hive from and make sure your bees have enough food and water inside the hive to last them 72 hours. If they don't have food and water to last them that long, you will need to feed them and give them water—see Feeding Your New Bees below.) After the three days are up, remove any wire or packaging from your hive, and you will soon see your bees start their **orientation flights**.

FEEDING YOUR NEW BEES

Most bees are ordered in the spring, just as the flowers are ready to bloom. This means they most likely ate all of their honey stores and are waking from their winter slumber. Much like any hibernating animal, bees are famished come spring. Honestly, a lot of bees are close to starving. It's necessary to feed your new bees right away when you bring them home.

Additionally, there are several other times throughout the year when you will need to feed them. We'll go over that in chapter 7; right now I want to focus on getting them comfortable in their new home. The feeder goes in front of the hive. What type of feeder you bought will determine exactly where you'll place it.

BASIC BEE FOOD RECIPE

You will find a million recipes for bee food. I'm going to share the most basic recipe with you. Whatever recipe you decide to use, though, make sure you use only natural ingredients. You can't feed synthetic ingredients or junk food to bees and expect them to produce a natural healing product—honey.

This recipe is measured in parts, so you can make as much or as little as you want. Parts express a ratio rather than a specific amount. For example, if you use two cups (two parts) of sugar, then add one cup (one part) of water.

Two parts organic cane sugar **One part purified or filtered water**

1. In a saucepan over low heat, mix together the sugar and water.
2. Stir until the sugar is completely dissolved.
3. Let the mixture cool.
4. Pour the sugar water (bee food) in your bee feeder.

 **POOL PARTY**

Did I mention bees *love* pools? Bees love pools more than a group of kids on a hot summer day. Other than completely enclosing your pool, you need to get used to the idea of sharing it with your bees. We used to keep a pool float tied to one corner of our pool, and our bees would use the rope to climb down to the water, hang out on the float, and just chill in the pool. Even if you don't want to have your own bee float, I suggest tying small ropes on the sides of the pool so the bees can crawl out. They aren't the best swimmers and will drown without a way out.

You will want to keep the bee feeder filled for several weeks after you get your bees, or until the nectar starts flowing in the flowers in your garden.

DON'T FORGET THE WATER!

Water is essential to all life, including a bee's life. Like many other creatures (including humans), a bee's body is made up mostly of water. So naturally, they need to drink water. Not only do bees use water for hydration, but they also use it to control the internal temperature of the hive. (To learn more about how bees use water to keep cool, see Bearding on page 105.)

Place a shallow water dish within a foot of the beehive and a couple of others throughout your property. You can put some rocks or marbles around the edges of your feeders and waterers, so the bees have a place to land. (If the water is too deep, your bees will drown.) Refill their water dishes daily with fresh water, just as you would do with any pet or livestock.

Chapter Six

THE BEE INSPECTOR

You are your own personal investigator and observer for your beehive. Think of hive inspections as looking in on your kids to make sure everything is okay when they're being too quiet. You don't want to be loud and let them know you're coming; you just want to take a peek to make sure everything is going as it should.

This chapter covers everything you need to know about hive inspections: when to inspect, how to inspect, what to look for, when and how to intervene. What type of hive you have will determine how often you perform inspections; however, much of this information is universal and can be applied to any beehive.

A good inspector is one who is neither seen nor heard, is careful not to disturb the investigation area, and takes good notes. As your hive develops and your bees start doing what they do best, your involvement will become less hands-on and more just observing.

TIMING IS EVERYTHING

While it's true that you *can* inspect your hive at any time of the day, it doesn't mean you *should*. The best time to snoop around in someone's house is when they're not there. So, when your bees leave to go to work, that's the best time for your inspection. There will still be plenty of bees inside the hive, but not all of them will be home.

Each hive is different, but pretty much all honeybees forage from sunup to sundown. Consider these their work hours. This would be the best time of the day to perform your inspections: a bit after the sun comes up until just before dusk on a clear, beautiful day.

HOW MANY BEES ARE BUZZING?

As I mentioned in chapter 4, when you order your bees, a two-pound package of bees will contain about 7,000 bees and a three-pound pack will be around 10,000 bees, give or take a dozen. A beehive in full swing, meaning at the top of production, will contain one queen and about 50,000 to 55,000 other family members. This number is drastically reduced in the winter when all of the drones have been kicked out on the street (see Say Goodbye, Romeo on page 107) and the queen slows down her egg production in preparation for the cold. A typical hive of 50,000+ in the summer will reduce to 20,000 to 30,000 in the winter.

Roughly 30 percent of your hive is comprised of foragers (female worker bees) that spend most of their lives outside of the hive. Of course, this number will vary depending on the climate and the availability of food. So, when you set out to inspect your hive, you are hoping that only about 35,000 are home.

TIMES YOU DON'T WANT TO INSPECT THE HIVE

» When it's dark, because all the bees will be at home (unless you think something is gravely wrong with the hive)

» When it's raining, because all the bees will be at home (unless you think something is gravely wrong with the hive)

» Early in the morning, because most of the bees will be at home

» In a high wind, because most of the bees will be at home, and when you take the cover off, they will be blown away

» When it's cold outside, because bees need their warmth and opening the hive will let the cold in and could actually kill them

WHAT IF THE QUEEN HASN'T BEEN RELEASED?

If you've opened the hive and noticed the queen hasn't been released, don't panic. There is still time for the colony to accept her. You don't want to release the queen too soon, because in doing so you could cause the colony not to accept her. In fact, if the bees don't release the queen, not only could they reject her, but they may end up killing her. This acceptance time is important to the colony. Once she is accepted by the colony, honey production may begin.

You can help the process along a little bit by adding a couple drops of water to the sugar plug on the end of the queen cage. This will soften the sugar a little bit and give the worker bees some help in removing the sugar plug. You may want to put a couple drops of water on the queen's cage as well, so she can get a drink.

Give the bees two more days and check on them again. By this time, the queen should be released. If your queen still hasn't been released by the end of the week, or if she dies in her cage, contact your bee supplier immediately about replacing the queen. A queenless colony will not survive.

WHEN TO TAKE YOUR FIRST PEEK

If you bought a pack of bees, or if your queen was packaged separate from your bees, you will want to inspect your hive within three days of installation to make sure the colony has released her. If you bought a nuc or full hive, it's better to wait one week before you do your first inspection.

HOW OFTEN SHOULD YOU INSPECT THE HIVE?

This is the big question. You will get a *lot* of variation in responses to this, so I'm just going to share with you what I do.

The first thing to remember is that when you inspect a hive, you are intruding in the bees' home. They won't always be happy to welcome you in. Say it's your cleaning day, you have your music jamming, and everyone is cleaning their respective rooms. You've got laundry going, dinner in the oven, you're pushing the vacuum, and then suddenly someone comes to the door and wants to come in for a visit. Are you happy to see them? Well, the bees are you, busy in their home doing what they do, and you are their uninvited guest.

IS MORE BETTER?

When you're first getting your beehive established, you will need to check on it a touch more often than you will once the bees are properly settled. As I mentioned earlier, if you bought a caged queen, you will need to check on her within the first three days. Once you know the queen is free and active, you can reduce your inspections to every two weeks for the first two months.

What about after that? When I was in my beekeeping class, my mentor told me to inspect the hive once a month, if needed, and never in the winter. That is the practice I've always followed. My approach to many things in life, including beekeeping, is if it ain't broke, don't fix it. I don't perform bee inspections just because the calendar says it's time to do so. Once I know my hive is established, meaning it has been a few months since the bees settled in to their new home, then I just observe from a distance.

Some beekeepers advocate performing hive inspections every week or every other week; while other beekeepers, like ones who have Warre hives, rarely inspect their hives.

I still haven't answered "how often?" My answer is once a month from spring to fall, *unless* the temperature drops below 40°F during the day. That's because bees **cluster** when it gets to be around 40°F, which means they gather in a clump in the middle of the hive to maintain their temperature. Opening the hive when they're trying to keep warm can give your bees a chill or possibly kill them. Of course, this is not guaranteed, but it could happen. There would need to be extenuating circumstances for you to open the hive when it's cold out—for instance, if they ran

cluster when bees huddle in the center of a hive to keep warm

out of food during the winter. If your bees are starving, by all means, open their hive and give them bee food.

TOO MUCH OF A GOOD THING

Inspecting your hive doesn't come without risk. Every time you open the hive, you are disturbing the bees' way of life. The possible negative effects of hive inspection include:

» Bee agitation

» Squishing the bees

» Inadvertently killing the queen

» Breaking their seal of protection (bees seal their hive with propolis to protect themselves from drafts and predators)

» Letting cold air in the hive and killing the bees

» Damaging the comb

» Hurting brood cells

» Inviting in predators

Please don't confuse inspection with observation. While it's true you are observing when you are inspecting the hive, it is completely possible to observe the hive without inspecting it. As a matter of fact, I encourage you to observe the hive daily. See Observing the Hive on page 86 for more.

So to summarize:

» Inspect three to five days after installing a new pack

» Inspect every two weeks for the first two months

» Inspect once a month until the fall

» Don't inspect in winter or after the outside temperature falls below 40°F, unless you suspect problems

OBSERVING THE HIVE

Observing your hive means you're just watching, sitting back, and relaxing as you take in all of what you are witnessing. A miracle at best. You're not opening your hive when you are observing.

When you are observing the hive, these are some of the things you should notice:

» Flight activity

» Smell: Do you smell a foul odor coming from the hive?

» Pests crawling in or out of the hive

» Dead bees by the entrance

» Bearding

» The hum of the hive: When bees are in the hive, they are in constant motion. The hum of the hive should sound like a small, quiet engine running.

» The weight of the hive: You do have to touch the hive by slowly lifting the back, but you don't have to open it. You want the hive to be heavier than when you first set it up.

bearding   when a large number of bees congregate on the front of the beehive, making it look like the hive has a beard; they do this to cool the interior of the hive

GETTING READY FOR A HIVE INSPECTION

Are you ready to witness the most well-oiled machine on the planet? From their communication skills to their honey production, honeybees truly are amazing insects. Now you have a front-row seat to this miracle. You get to observe the inner workings of these amazing creatures, watch how each of the different roles is performed, witness baby bees developing, and see honey being made—not to mention, the precise construction with which the bees build their comb is a sight that will never grow old.

BEE CALM

Here's the thing: You are going to be scared at first when working with your bees. You may even be a little bit terrified—especially if you're allergic, like me. But here's the tricky part: You need to stay calm. Just as dogs can sense the tension in the air, your bees will react to your mood. I know suiting up like you're going into an infectious disease lab and surrounding yourself with thousands and thousands of bees that have the ability to take your life is a bit intimidating. No worries, right? Despite everything I just said, I want you to try to remain calm. Doing yoga on a sunny beach with a light breeze calm.

Increased breathing, sudden sharp movements, or anxious behavior can send the wrong signals to your bees and put them on edge. So think happy thoughts, calm your breathing, move slowly and intentionally. Before you know it, you and your bees will be very comfortable around one another. In fact, in time you might even enjoy one another's company.

NEVER FROM THE FRONT, UNLESS YOU WANT A FIGHT

When performing hive inspections, it is important to approach your hive from the side or the back, but never from the front. This little tidbit of information is good to remember when you are choosing your hive placement. Make sure you'll have enough room around your hive to access it from all sides.

Bees establish flight paths to and from their home. They know where they are going and how to get there. If you are standing in their way, you've now rerouted their GPS. In addition to taking them off course, you may be mistaken for a predator trying to attack their hive, causing security bees to take proactive measures. Remember, their entire mission in life is to protect their hive, and they will give their lives to do so. Approaching your beehive from the side or back will reduce the chance of you getting stung, and you won't interrupt the bees' daily commute.

ALL ABOUT YOUR SMOKER

A smoker is an essential part of a beekeeper's tool kit. The smoker provides protection for you and aids in inspections.

A bee smoker works in two ways. First, it helps mask the warning scent bees emit when they feel there is danger and want to sound the alarm. Second, where there's smoke, there's fire. When bees see smoke, they think there is fire. If the bees were in their natural habitat of the forest and there was a fire, they would start eating large amounts of honey in preparation for losing their home and starting over. Just like humans, when bees eat a large meal they become lethargic and lazy. So, when you're smoking, the bees are more concerned about losing their hive than paying attention to what you're doing.

While the smoker isn't always necessary, it is important to always have it lit before you begin your inspections. It takes time to get the smoker lit, and it's better to have it ready and not need it than to need it and not have it ready.

HOW THE SMOKER WORKS

A bee smoker is basically a can where you make the fire, a bellows to keep the fire going, and a spout where the smoke comes out. You put some fuel in the can and build a little fire, then you pump the bellow to push oxygen through the fire, and the smoke comes out of the spout.

You'll need something to get the fire started, like a piece of cardboard, newspaper, or a pine cone. Use a match to light that, then toss the started fire in the bottom of the can and start adding light, thin kindling material. Wood shavings, pine needles, dried leaves, and straw are some of the possibilities for kindling. But they all burn up quickly, and you'll need something that will keep burning longer. Small twigs or wood chips are good choices. Just make sure you don't pack them too tightly, or you won't be able to draw air through your smoker and the fire will go out. Pump the bellows after each addition to help the new material catch fire.

Whatever fuel you use, make sure it's 100 percent natural and untreated. Whatever was in the fuel will end up in the smoke, and you don't want your bees (or you!) inhaling any chemicals.

THE INSPECTION PROCEDURE

The main goal during inspection is to finish up everything in about 10 minutes. Any longer and you could expose the hive to too much cold or wind, or simply overstay your welcome. So have your inspection checklist and all of your tools ready and close by, so you can get in and get out as quickly as possible.

Here's your preinspection checklist.

» Inspection checklist

» Safety gear

» Hive tools

» Beekeeping journal (and pen)

» Smoker lit and ready to use

All right then, let's dive in.

 ## TO SMOKE OR NOT TO SMOKE?

Even if you choose to use a smoker, it may not be necessary to use one every time you do an inspection. As you mature in your beekeeping experience and become comfortable handling your bees, you may not see the need for the smoker every time. In addition to becoming comfortable, beekeepers who prefer the more natural method of beekeeping prefer not to use a smoker at all.

Personally, I fall into this latter category. While we have a smoker, and will use it if necessary, I haven't found the need in several years.

WHAT TO LOOK FOR

When you are inspecting your hive, there are some specific things you should look for. The more you observe your hive, the better you will get at recognizing everything. It will take time to become familiar with all of the components of your beehive, so don't be discouraged if you don't see everything listed here the first time you inspect.

» **Identify and locate the queen:** If you didn't buy a queen that was marked, refer to page 16 where I go over the physical attributes of the queen, to help you recognize her.

» **Comb development:** If you are using frames with foundations, your bees should have started to draw out, or build out the combs from the foundation. If you used foundationless frames, your bees should be building their comb.

» **Eggs:** Bee eggs will look similar to tiny specks of rice, and may be hard to see. Each cell should contain only one egg.

» **Larvae:** Larvae are the next step in bee metamorphosis. At this stage, they will resemble white maggots. Some of the larvae cells will be capped and some uncapped, meaning some will be sealed and some will be open.

» **Pupae:** These are bees in the final stage of development, the pupa stage. At this stage, they will start to resemble bees. They have wings, eyes, and legs. At the end of the pupa stage, which lasts 7 to 14 days, the baby female bees will chew their way out of the cells.

» **Brood:** Collectively, all the developing bee cells are called brood. You should see drone cells, worker cells, and queen cells (if queen cells are needed by the colony) in various stages of bee metamorphosis. The worker cells will generally be located in the center of the frames and be almost flat—unlike the drone cells, which will have a domed cap. The queen cells will be larger than any other.

» **Honey production:** Honey cells will be filled with shiny nectar, and the caps will be more translucent than those on the brood cells.

WORKER BROOD CELLS

QUEEN BROOD CELLS

DRONE BROOD CELLS

» **Filled comb**: Whether by brood or honey, your combs will start to grow or your frames will fill. If three-quarters of your frames are filled, then it is time to add more room for growing. In Langstroth or Warre hives, this would mean it is time to add another box.

» **Disease and pests**: I'll discuss what you might see (and what to do about it) in chapter 9.

» **Dead bees**: A few dead bees is not a big deal. But do you see an abundance of dead bees in the hive that aren't being removed? If so, then you will need to inspect further into your hive for signs of pest or disease. Other reasons for an excess of dead bees are extreme cold, exposure to poison, or starvation.

FIVE THINGS EVERY HIVE SHOULD HAVE

When performing hive inspections, regardless of the type of hive you have, you should see these five things.

1. Bees: workers, drones, and the queen
2. Brood cells: baby bees in the making
3. Honey cells
4. Propolis: bee glue
5. Comb: beeswax

HONEY CELLS

BROOD CELLS

PROPOLIS

EMPTY CELLS

LANGSTROTH HIVE INSPECTION

In a Langstroth hive, you will have the bottom board, the boxes, the inner cover, and the top board.

» **Bottom board**: The bottom board will have the entrance into the hive. Some beekeepers like to add a wire bottom so that small hive beetles will fall through the bottom onto a sticky board, keeping the bees safe inside.

» **Boxes**: The boxes are the main body of your hive and come in three different sizes: shallow, medium, and deep. The deeps are generally used for brood production, while the shallows and mediums are used for honey production. (Although the bees don't know that, and will use the boxes as they like.)

» **Top**: You will have an inner cover and a top board (outer cover) on the top of your hive. The inner cover helps prevent the bees from gluing the top board to the boxes with their propolis.

» **Queen excluder**: A queen excluder looks like a grate and is installed between the deep box and the medium or shallow box. The spaces are big enough to let the worker bees through, but too small for the queen bee to climb through. Beekeepers do this to prevent the queen bee from laying eggs in the honey boxes. This also keeps the queen in the bottom of the hive. (Side note: queen excluders are not included with the purchase of a Langstroth Hive and will cost additional money. Some beekeepers choose not to use them because they do not want to restrict the queen's movement within the hive. It is a personal choice to use one or not.)

LANGSTROTH INSPECTION STEPS

1. The first thing you will want to do is lightly smoke the front of your beehive if you plan to use a smoker.

2. Remove the top board. If the top board is stuck, you will need to use your hive tool to remove it.
3. Lightly smoke the inside of the hive again.
4. Remove the inner cover and place it off the ground but within reach.
5. Depending on how many boxes you have, you will want to repeat this process for each box.
6. Carefully remove a frame to inspect it. Look for the items on the inspection checklist. In my opinion, it is not necessary to inspect each frame; just a couple will do.
7. Replace each frame after inspection, paying the utmost attention to the bees. Unfortunately, many queens have met an early death from being squished during a hive inspection.
8. Replace the inner cover, then the top board.
9. When you are finished, slowly move far enough away from your hive so you can safely remove your bee suit without a slew of bees covering you.

TOP BAR HIVE INSPECTION

A top bar hive inspection is similar to a Langstroth hive inspection; you will perform the same steps, with the exception of removing the inner cover and boxes. The main difference between the Langstroth and top bar hive inspections is the handling and care of the frames.

Top bar frames are not as structurally sound as Langstroth frames, and the comb can easily break away from the top bar due to its weight. This is especially true if the comb is filled with honey. Use extra caution when handling the frames for inspection.

WARRE HIVE INSPECTION

Keeping true to the Warre principle, the hive was not intended for weekly inspections—or even monthly inspections, for that matter. However, some state laws may require you to do inspections at regular intervals.

One of the main reasons for hive inspection while using a Warre hive is to see if your bees need a new box. Generally, you will have to add a box to the bottom in the spring and again in the summer. Take the same extra care required for the top bar hive when handling the frames, as they lack support and are very fragile.

TO INSPECT YOUR WARRE HIVE:

1. Lightly smoke the entrance of your beehive, if you plan to use a smoker.
2. Remove the top board.
3. Lightly smoke the inside of the hive again.
4. Proceed from step 6 of the Langstroth hive inspection.

START YOUR BEEKEEPING JOURNAL

Recordkeeping is an essential part of beekeeping and animal husbandry. There's a saying I like to apply to endeavors like this: "You don't know where you're going unless you know where you've been." Meaning, you can learn a lot about your present by looking at your past. Journals promote good beekeeping practices.

Have a dedicated journal just for your beekeeping records, and update it weekly. If you have more than one hive, keep a separate journal for each hive. We number each hive and keep a journal with the number on the front of it.

Some of the things you will write in your journal include:

» A list of all expenses, which will be important for tax time should you decide to sell honey or bee-related products.

» The health of your hive when you inspect it.

» Anything that concerns you or things that seem abnormal.

» When you did your last inspection.

» If you treat for pests, what you used and how much you applied.

In addition to keeping notes and expenses in your journal, you can also add an inspection checklist (see below). A checklist is a great idea for first-time beekeepers. You're going to have so much going on in your head that having a checklist is important, until you get into the swing of things.

A preinspection checklist (page 89) can also come in handy. For example, you may open the hive and then realize you forgot your hive tool or brush, so you put the top back on and then go to get your tool. This

INSPECTION CHECKLIST

The key to inspection is to have a specific purpose. Think about what you are looking for when you do your inspection. Remember, an efficient inspection should take no longer than 10 minutes. At the end of each inspection, make sure to go back to your journal, fill in the checklist, and add notes about all of your observations.

» Check on the queen. Can you identify her?

» Is the queen laying eggs?

» Are the bees making new comb?

» Are the bees making honey stores?

» Is the honey ready to extract? (See chapter 8 for signs that the honey is ready to extract.)

» Do you see brood?

» Are new queen cells forming? (See page 17 for an image of a queen cell.)

» Is there a funny odor?

» Do they look like they may swarm? (See chapter 10 for what those signs are.)

» Do they need more room?

» Are there any signs of disease or pests? (See chapter 9 for what those signs are.)

back-and-forth disturbance could tick off a couple of the guard bees. A checklist can help you stay on track and remain organized until you've performed enough inspections that they become second nature.

THERE'S AN APP FOR THAT

In a world of technology, beekeeping hasn't been completely left behind. I have many commercial beekeeping friends who swear by the HiveTracks beekeeping app. It's the most popular beekeeping app at this time. Although the free version gets horrible reviews, my friends swear by the paid version. HiveTracks can be your beekeeping journal, but without the pen and paper. For other app ideas, see the Resources section of this book.

BE PREPARED FOR BEE STINGS

We all know it's coming, so better brace yourself now for the inevitable. You will get stung. I can't say where or when, but sooner or later it will happen. When it does—and it will—it's best to be prepared.

As I mentioned in chapter 2, drones (male bees) don't have stingers. The queen has a stinger, but she reserves its use for defeating her possible successor. If you get stung, a worker bee did it. Worker bees are the only ones who will sting you.

When a bee stings you, whether by accident (you placed your hand on her and oops!) or she was defending her colony, she will die. Her stinger is attached to parts of her abdomen, so that when she stings you, her insides are literally ripped out. Between the two of you, the bee gets the worse end of the deal by far.

FIRST AID FOR BEE STINGS

There is venom in a bee's stinger. The longer the stinger stays in you, the more the venom will spread. The first thing to do when you get stung is to remove the stinger from your skin.

You can remove the stinger by scraping it out with a straight edge, such as the edge of a driver's license or a credit card. I don't advise using tweezers, as they could release more of the venom in your skin.

Bees release a pheromone when they sting you to alert other bees that you are a threat and to come attack. So wash with soap and water as soon as you remove the stinger.

If you are allergic to bee stings, experiencing extreme shortness of breath, extreme swelling, hives, dizziness, or nausea, seek immediate medical attention. Do not try to treat the sting yourself. Just try to remove the stinger and get to a medical facility as fast as you can.

OLD-TIMEY REMEDIES FOR BEE STINGS

Old-timey remedies are often tried and true. However, sometimes they are about as reliable as the weatherman—you have a 50 percent chance of them actually working. The venom in bee stings is acidic, so anything that helps neutralize this acid should help relieve the pain. After you have removed the stinger and washed your skin, try one of these traditional remedies.

» **Baking soda**: Make a paste with baking soda and water, apply to the sting, and let sit until dry.

» **Toothpaste**: Put a dab of toothpaste on the sting until the pain subsides.

» **Tobacco**: Wet the tobacco and place it on the sting until the pain goes away.

- » **Copper penny**: Tape the copper penny over the bee sting for 15 minutes.

- » **Tea bag**: Wet a tea bag and press it on the bee sting until the pain subsides.

- » **Meat tenderizer**: Mix one part meat tenderizer to three parts water and place on the sting for 30 minutes.

- » **Aspirin**: Crush an aspirin and mix it with equal parts water, place on the sting, and let dry.

NATURAL REMEDIES FOR BEE STINGS

- » **Honey**: Place raw, organic honey on the sting and cover with a bandage; rinse after 30 minutes.

- » **Plantain**: We're referring to the common weed here—not the large, banana-like fruit. Crush some plantain (the leaves) to make a poultice and secure on the sting with some tape.

- » **Essential oils**: Diluted oils such as lavender and tea tree oil have been used to help relieve the pain of bee stings.

- » **Apple cider vinegar**: Soak a cotton ball in vinegar and place on the sting until the pain dissipates.

Chapter Seven

THE BEEKEEPING YEAR

Your entire first year of beekeeping will be a learning curve. It's a year of firsts for you and your bees. You may experience your first sting, but you may also experience your first home-grown honey. Think of this chapter as a season-by-season checklist for everything that should be happening. After your first year you'll be a pro at beekeeping and will know what you'll need to do each season, from winter feeding to spring honey flow.

Most beekeepers order their bees for spring delivery. However, depending on the climate where you live, you may be able to buy them any time of the year. Regardless of the season or when you bring your bees home, this seasonal almanac will remain the same. For the sake of saving paper, and your time, I'm going to write this bee-keeping year as if you received your bees in the spring, and you can adjust accordingly, if necessary.

THE FIRST TWO MONTHS

Let's say you just installed your new bees and the flowers are starting to bloom. You will now need to follow all my instructions in chapters 5 and 6 for the first two months. You will need to keep the feeder full with the bee food recipe on page 78 and make sure your bees have a fresh supply of water every day.

During the first two months, you will need to check on your bees daily to observe their behavior. I'm talking here about the outside-the-hive observations I outlined on page 86. In addition to daily observation, you should perform your very first inspection three to five days after installing your bees, to make sure the queen has been released if you bought a pack. After you've observed that the colony has accepted their queen, you will inspect every other week for about a month. After your initial couple of hive inspections, it's good to leave the bees alone to do their thing during the nectar flow in the spring.

SPRING, THE FIRST YEAR

As a beekeeper, it is essential to learn about your local nectar flows. You can contact your local county extension, beekeepers association, or garden club for this information. The nectar flow will determine when you need to feed your bees (when there isn't much nectar) and when you need to hold back (when there's plenty) so they'll forage for their food.

Your bees are new to your area this spring. They will take a lot of orientation flights to familiarize themselves with the local forage and water supplies. This is why it's a good idea to have plants in your yard for your pollinators. In case you missed it, I've included a list of plants bees love in chapter 4.

NO HONEY FOR YOU

If you got into beekeeping strictly for the liquid gold called honey, I hope you are a patient person. Chances are you won't harvest your first honey until spring next year. Your bees will need time to establish their colony, build comb, and store honey for their survival this year. There is a possibility of a fall harvest, if you have a good nectar flow this spring and summer. Realistically, though, don't plan on harvesting honey for at least one year. Trust me, it's worth the wait.

SPRING, THE SECOND YEAR

Spring is the most active season for bees. As they start to venture out from their winter slumber, you will notice the bees returning to the hive with yellow legs. That's pollen being carried in the **pollen baskets** on their back legs. Inside the hive, the queen is busy laying eggs and the workers are busy raising brood. Once the nectar flow starts and you see your bees gathering pollen, you can

pollen basket an area on the hind legs of honeybees where they collect pollen while foraging to transport it back to the hive; hairs hold the nectar-moistened pollen in place

stop feeding them. But you should always supply fresh water regardless of nectar flow or time of year.

APRIL SHOWERS BRING PROBLEMS

While rain may be good for your gardens and plants, it's not good for bees. Bees do not forage in the rain. It's extremely important to pay attention to the rainfall in the spring because the bees will be very hungry and have a lot of brood to feed. Not only will bees not forage in the rain, but rain washes away pollen and dilutes nectar. Meaning, even after the rain is over it takes a while for the plants to recover enough to provide the bees with everything they need.

If you're experiencing a particularly rainy season, no matter what time of year, you should provide your bees with some bee food so they don't go hungry.

CLEANING HOUSE

I can't imagine any beekeeper who wants to walk out to visit their hive and see a slew of dead bees, especially if you're new. The first thought that comes to mind is that they're all dead. If you recall the science lesson about the life span of bees in chapter 2, you know the bee has a very short time on this earth. Bees do live longer in the winter than in the summer, but even so, they can't live forever. By the time spring rolls around many of your bees, with the exception of the queen, are at the end of their life cycle.

Once the temperature starts to warm up, your worker bees get busy cleaning house and removing all the bees that died over the winter. This can be an alarming number, but rest assured, your bees should be just fine.

If you notice the hive entrance gets clogged with dead bees, you may want to help by removing them with your bee brush. However, your worker bees should be able to handle this, given enough time. Just keep an eye out—observe.

🐝 DEAD BEES

During the winter, the bees will cluster in the center of the hive to stay warm. Kind of like the human equivalent of snuggling when we're cold. During this clustering time, all other household chores are put on hold, including taking out the trash—aka dead bees. When the temperatures start to get warmer, everyone gets back to work and there is a lot of spring cleaning to do.

Some of the various reason you will see dead bees outside your hive when the temperature starts warming up are:

» **Age:** it was just their time to go

» **Cold:** bees will die in the winter from cold; generally, these are the older bees that were on the outside of the cluster

» **Wet:** too much moisture, which leads to cold

» **Starvation:** not enough winter food stores

» **Disease**

» **Pests**

SPRING SWARMS

When bees become crowded in the spring because their colony has grown, or if the workers made new queen cells and have crowned a new sovereign, it's time to swarm. A swarm is when half of your colony and the old queen leave the hive in search of a new home. The remainder of your bees and the new queen will work together on building their colony.

Beekeepers have a love-hate relationship with swarms. They love to catch swarms to build their apiary. This is one way to get free bees and expand your hives. However, they hate it if they either can't catch the swarm, or if they didn't plan on expanding. In which case, they will lose half their bees.

Losing half the workers does leave the new hive vulnerable and a little weak. Sometimes adding an entrance reducer to the hive after a swarm splits helps the new colony defend its entrance while they build their numbers back up. (See page 70 for a recap on entrance reducers and page 144 for more on swarming.)

HONEY FLOW IN THE SPRING

Spring is the season all beekeepers look forward to. Late spring is when the honey starts flowing and their hives start growing. Spring honey is a favorite among most people because of the fragrant hints the honey has from the spring flowers. This is especially true if your bees have fruit trees or bushes to forage on.

My friend's blueberry farm produces the best-tasting spring honey I've ever had. If your bees are bursting at the seams with honey stores, now is the time to get your honey extraction equipment ready. Just remember to leave some for the bees; they need it to survive.

SUMMER

During the summer months, the bees continue to forage all morning, noon, and even into the night. Since the days are longer, the bees have more time to gather pollen, nectar, and water. Your colony will be at its greatest population during the summer months, with each member preparing for the coming winter.

If you haven't already done so, now is the time to remove your entrance reducer. Failing to do so could cause your hive to overheat. The summer heat will also cause the bees to be less active, and they will beard on the outside of the hive to stay cool.

BEARDING

One thing you'll notice that may seem weird during the dog days of summer is bearding. Bearding is when a lot of your bees hang out on the outside of their hive, in a formation that resembles a man's beard. This bearding during the heat is different than the bearding you will see when your bees are getting ready to swarm.

Have you ever seen wax melt in the heat? Like when your lip balm gets too hot when you accidentally left it in the car? The same thing happens to beeswax inside the hive during the heat of the summer. To keep the hive cool,

BEARDING

many of the bees will leave the hive and hang out on the outside.

Another pretty amazing way bees stay cool is by evaporation, creating their own air-conditioning. They do this by collecting water and adding it to the comb. Then they fan the water with their wings, causing it to evaporate and cool down the hive. Pretty neat! This bee-conditioning is another major reason why you need to provide your bees with a constant supply of fresh water. In addition to water, you can help your hive stay cool by making sure it's located in a shady spot during the summer.

ORIENTATION FLIGHTS

Do you remember your job orientation? On your first day, someone shows you where the break room is and where to park. When it's time for the bees to leave their hive, they will take an orientation flight. I've always enjoyed orientation flights because we get hundreds of bees just flying all over our property pinging into everything. Don't be alarmed if you see a bunch of bees flying in places you haven't noticed bees before. These are the new recruits getting familiar with their surroundings during orientation flights.

Another thing you may notice during orientation flights is it is lightly raining when there isn't a rain cloud in the sky. Hint, hint, that's not rain falling from the sky, so don't look up. The light sprinkle you may feel is actually bee pee. Yep; people do it, birds do it, and bees do it. Oh, and those mustard-brown streaks on the front of your once-spotless beehive? Take a guess.

FALL

Fall is the bees' last hurrah to gather their harvest for the long winter. This is the last nectar and honey flow for the season. As the summer plants go to seed, bees take advantage of this to make more honey to store for the cold winter.

Fall is another time you need to pay attention to the flowers in bloom, the rainy days, and your bees' activity. Are they still returning to their hive with yellow legs? If your bees don't have enough to forage on during the fall, they will eat their winter stores. Which means they will starve in the coming months if they're not fed. Best to feed them now so they can save their honey for the winter.

SAY GOODBYE, ROMEO

Say goodbye, Romeo, the queen has no more use for you guys. Spring and summer are the mating seasons for bees. In the fall and winter, the queen takes a break. The queen doesn't start laying eggs again until the end of winter, so there will be a new group of bees emerging in the spring to replace the old ones.

Unfortunately for the drones, the queen has no further use for them. No hive likes deadweight, nor do they have extra honey stores to share with bees that aren't serving a function. When the leaves start to turn brown and the air gets chilly, the drones get their eviction notice and death sentence—without the happy ending that the summer and spring drones got.

In the fall, you may see the worker bees taking out the deadweight, some even before total development. If you notice dead bees by the front door of your beehive, take note if they have a stinger or not. If they don't, you'll know it was the men getting the fall boot.

THAT FALL SMELL

Remember when I talked about plants flavoring honey in chapter 4? Not only do plant pollen and nectar help flavor the honey, but they also give it a scent. Needless to say, the first time I smelled the fall honey production, I thought something had died and started rotting in the backyard. Yeah, it's that nasty. However, I have friends who live up north who adore the flavor of fall honey. They swear by the taste and claim that people seek out their fall honey. I'll take their word for it.

Typically, that unique fall smell can be attributed to goldenrod and other native plants from the aster family. I'm not sure whether this is the grand design, but many beekeepers prefer to let their bees keep their fall honey for this reason—including this beekeeper. Well, that and for other reasons I'll talk about in just a moment.

Other than stinky honey, dead or soon to be dead drones, and possible starvation, your fall beekeeping schedule should be pretty uneventful. If you plan on a fall inspection, make sure to do it before the weather starts to get too cold—remember the rule is not to inspect if the temperatures are below 40°F. Your bees will be busy sealing all the nooks and crannies of their home with

HONEY FLOW IN THE FALL

There are two times a year that beekeepers harvest honey: spring and fall. Each season has its own benefits and risks. Generally, beehives will have more than enough honey to share with you and others, but sometimes they have just enough for their own colony to survive. This is where you, their beekeeper, have to be diligent about your observations and keeping a journal. The biggest concern about harvesting fall honey is not to take too much, and to not squish the queen in the process.

If you follow the weight suggestions I mention in How Much Honey Will Your Bees Need for the Winter? (see page 111) and you are very careful not to kill the queen, you can reap the rewards of your bees' fall honey production. With a strong colony and a productive summer, your bees should produce enough honey to keep them fed all winter and have more than enough for you to enjoy.

propolis to keep it buttoned up in the winter. The last thing you want to do is strip their house of insulation because you wanted to take a peek inside.

WINTER BEE FOOD

Because liquid bee food can freeze in the winter, you will need to give your bees something that will be accessible in the cold, such as organic raw honey, candy boards, bee cakes, or bee patties. You place them in the top of your beehive in late fall, before the winter cold hits and before the bees seal their hives for the winter.

Candy boards, bee cakes, and bee patties are all pretty big, to last the bees through the winter. They are solid sugar food for the bees if they run out of honey. You can make them yourself—there are plenty of recipes online. It's much like making regular candy, except easier, and you use ingredients from your kitchen. If you don't feel like Betty Crocker, you can buy them already made online from bee suppliers or in stores that sell beekeeping supplies.

WINTER

Winter is a pretty boring time for the beekeeper. All of the action is taking place inside the beehive, and there is very little outside for you to see. Your bees have one main job to do in the winter and that is to keep the queen bee alive and well. All other production comes to a halt. Now it's all about the queen and survival of the fittest.

Bees stay in their hive during the winter to stay warm. As the temperature dips into the low 50s, your bees will start to gravitate toward the center of their hive and form a cluster. This cluster will expand and contract with the fluctuation of the temperature. It will expand when the temperature is warmer and contract as it gets colder.

The queen will be at the center of the cluster, with the older bees on the outer edges. As the bees consume their honey, they move their cluster throughout the hive. The bees rely on their honey stores to keep them alive through the winter. By the time spring rolls around and the nectar starts to flow, many bees are close to starving to death because they've eaten through all of their honey stores.

One way to check on your bees in the winter without opening their hive is to lift the back of the hive. Is it heavy or light? If it's light, you need to give them food; the lack of weight means there is a lack of honey. Another way to check on them in the winter is to put your ear close to the hive. An active hive hums like a small engine running. If you don't hear a sound coming from your hive, this could be cause for alarm.

HOW MUCH HONEY WILL YOUR BEES NEED FOR THE WINTER?

This is a loaded question, and one that doesn't have a definitive answer. Your bees may require anywhere from 30 pounds to 90 pounds of honey to last them through the winter. The colder your climate, the more honey they will need. In places like Maine, the bees will need a minimum of 90 pounds to make it through the winter. In warmer climates like Florida, they will likely only need 30 pounds of honey stores. Knowing how much honey your bees need to live will help you determine how much honey you can harvest in the fall, if any.

While it is not a great idea to unseal the hive in the winter, it's an even worse idea to let your bees starve to death. If you lift your hive in the winter and it feels light, you will need to add food to save your bees (see Winter Bee Food on page 109).

MOISTURE KILLS

Most bees can tolerate the cold, but not moisture combined with cold. Excessive moisture in a hive will kill a colony as quick as any disease or pest. Bees will create condensation from their body heat when it meets the cold walls of their hive. This condensation makes moisture inside the hive. If the airflow is restricted, there is no way for the moisture to escape and it will wet the bees. Wet bees + cold = death.

Here are some ways to help reduce moisture in a beehive in the winter.

» Don't restrict airflow into the hive. But do protect the hive from strong winter winds.

» Add an absorption blanket (aka moisture quilt) inside the inner cover. Moisture quilts can be made from various materials. A basic example is a mesh bag filled with cedar chips and installed in the top of the hive where condensation tends to collect.

» Place newspaper, cardboard, or other absorbent material in an empty box on top of the hive.

RELAX

Are you reading all of this and thinking, "Wait a minute, you said bees are the easiest livestock you ever raised!" They are. Your first year will be the hardest as far as the learning curve goes, but you'll soon come to enjoy the lack of hands-on attention your bees require. They are not as fragile as they seem. I mean, they have been around for thousands of years. You and your bees will be just fine.

Chapter Eight

ALL ABOUT HONEY

This is the chapter we've all been waiting for—or at least I have. You will learn all about honey, from its production to its harvest. Oh, what a sweet reward! Something this decadent takes time to make, so keep in mind that beekeeping is a journey. It shouldn't be looked at as a race or a hobby that provides instant gratification. To me, honey is a bee's way of saying thank you for taking care of it for a year, protecting it, and providing it with food and water when it needed it. Honey is a sweet golden gift given as a token of appreciation by a bee.

FIRST THINGS FIRST: WHAT IS HONEY?

Simply put, honey is processed and cured nectar. Bees rely heavily on honey as their only food source when pollen and nectar are not available. Some people may refer to honey as "bee vomit." Are those people right? Well, sort of, but not exactly. Bees are equipped with a sac at the bottom of their throat called a **honey sac**. Honey is the result of bees processing nectar in the honey sac and infusing enzymes into it. They store the nectar in their honey sac until they get back to the hive.

> **honey sac** a gland at the end of a bee's esophagus in which it stores nectar gathered while foraging

HOW BEES MAKE HONEY

Thousands of worker bees go out to forage for both nectar and pollen. They consume nectar from plants and flowers and store this nectar in their honey sac, where it mixes with enzymes. From there, they fly back to the hive and transfer their nectar to another worker bee, called a house bee. The house bee then repeats the same process, but transfers the nectar into the comb cells, where it becomes honey.

The bees fan the honey to help the liquid evaporate, similar to the way you would reduce a liquid in cooking to make it thicker. At this point the honey is still green, which means it hasn't been processed enough by the bees for consumption and could easily spoil. When they've reduced the liquid enough they cap it (seal it with beeswax) for long-term storage. Once it is capped, it is honey.

IT TAKES A LOT OF BEES

The term "being worked to death" brings on a whole new meaning with bees. I know some of you may feel like you'll work until you drop dead, but bees actually do that. From the moment they crawl out of their cell, they work until death for the greater purpose of their hive. No poolside honeycombs, no beehive retirement community, just 30-plus days on this earth of constant labor.

Bees will travel up to five miles from home in search of plants to forage on. If humans had to walk the equivalent distance to gather food as bees do, most would starve. Remember this when you're thinking about what your bees may be exposed to. You can have the best organic garden around, but your bees will still forage at the nearest chemical plant. That is why it is almost impossible to have truly organic honey. Not every plant the bees visit will be free of toxic chemicals.

One bee only produces about 1/12 of a teaspoon of honey during its lifetime. With an average beehive consisting of 55,000 bees, they will collectively need to fly around 55,000 miles just to make one pound of honey! Let that sink in for a minute.

When you buy a pound of honey at the grocery store, it took about 770 bees to make that, along with nearly 55,000 miles of travel. How many flowers does this mean they had to visit to fill the plastic bear you just bought? More than 2 million. This is just for one pound of honey. Do you understand why we call this stuff liquid gold?

As you read this and ponder the magnitude of what it means, you need to reflect on the importance of plants in our lives and for bee survival. We need plants, plants need

HOW MUCH HONEY WILL ONE HIVE PRODUCE?

An active beehive can produce more than 200 pounds of honey in a year. For argument's sake, let's talk about Langstroth hives, since they are the most common. In chapter 3, I mentioned the different parts of the Langstroth hive, including the deep, medium, and shallow boxes. The deep box, filled with honey, could weigh around 70 pounds, the medium box, when filled with honey, will weigh about 50 pounds, while the short box filled with honey will weigh around 40 pounds.

As your colony grows and honey production flows, you will add boxes to your hive—in some cases as many as five or six boxes, all containing honey or brood. Remember, I said way back in chapter 1 that beekeeping can be hard on the back and you will need a partner to help you. The sheer weight of a full hive can be too much for one person to handle. This is why I strongly suggest you enlist some help. Free honey can always be motivation to help a beekeeper

lift heavy beehives on harvest day.

Of course, these numbers are just an example. Many variables will factor into your true honey production, such as the rain, the forage available, and the health of your colony. When drooling over honey daydreams about all the liquid gold you will be swimming in, remember how much honey the bees will need to keep for themselves to survive the winter (see page 111).

bees, and bees need plants. It's the circle of life. I always say beekeepers and gardeners go hand in hand; one cannot thrive without the other. This is why I mentioned in chapter 4 that you should plant a bee-friendly garden. Include some forage for your bees; not only will your plants feed your bees, but what you plant can help flavor your honey. As an added bonus, you can also harvest whatever fruits, vegetables, and flowers you plant. It's a win-win.

THE HONEY HARVEST

Since I already burst your bubble in chapter 7 about waiting to harvest your honey during your first year of beekeeping, this information won't sting as bad. As a new beekeeper, you probably shouldn't harvest honey until your second year of beekeeping. The main reason is so your colony can get established, build their comb, increase their numbers, focus on their health, and build up their honey stores. However, in that second year all bets are off, honey. As a matter of fact, if you're not selling your honey, you'll probably have it dripping out your ears.

BEST TIME OF YEAR TO HARVEST HONEY

Truth be told, the best time to harvest honey is when your colony tells you to. What do I mean by that? I can sit here and tell you to harvest honey at the end of June and again at the end of September, but in all honesty, if your hive isn't ready, it's not ready.

Your monthly inspections will tell you when it is time to harvest honey. I should remind you that the one time you should not harvest honey is in the cold winter. That is a no-no. Have you ever tipped a honey bottle upside down in the cold? You're kind of going nowhere fast with that one. Cold honey doesn't extract well; glaciers move faster than cold honey.

My general rule about when to extract honey is to pay attention to the nectar flow. Watch your honey stores in your hive at the end of each nectar flow. That is when your honey supply will be at its max and you will know it's time to harvest.

WHEN IS YOUR HONEY READY TO HARVEST?

Your honey is ready to harvest when you perform your hive inspection and most of the frames in your hive box have honey. We talked about the difference between brood cells and honey cells in chapter 6. Once you're able to identify the honey frames, you will need to check to see how many of the honey cells are capped. Capped honey cells are ready for harvest and uncapped honey cells generally have too much moisture in them to harvest—which is why the bees haven't capped them yet.

When the bees collect nectar, it has around an 80 percent moisture level. As they process it and reduce the moisture by fanning, they will bring the moisture level down to around 18.6 percent, then cap it with the beeswax to seal and protect it. When your frames are about 80 percent capped, you know it is time to extract the honey.

🐝 NECTAR FLOW

Nectar flow is also called honey flow because it is the time of year the bees are producing a lot of honey. Most experienced beekeepers and beekeepers associations will be able to tell you when the local nectar flow is in your area. The main honey or nectar flows will last several weeks, while most of the local flowers are in bloom. That is why spring is the main honey harvest time for beekeepers, followed by fall.

Weather can greatly affect nectar flow, especially if you've had a lot of rain. Instead of one huge rush of nectar flow, it may be a long and slow trickle.

The local nectar flow is another important thing to add to your beekeeping journal.

Observe what is in bloom and write down what month it is and how long the flowers are in bloom. Take note of all the different flowers that are in bloom during each season, and if you extracted honey, what it tasted like. You will definitely notice how the different flowers flavor your honey.

UNCAPPED HONEY

Uncapped honey cells will have too much moisture in them and can be prone to fermenting. However, sometimes bees are so busy collecting nectar and making honey that they simply don't have the bee-power to cap all the honey. In this case, some of the uncapped cells are in fact ready to harvest. You can buy a honey refractometer to measure the moisture levels in your honey cells. The refractometer helps take all the guesswork out of knowing whether the moisture level is right. If your honey reads a moisture level of 18.6 percent or lower, it is ready for harvest.

UNCAPPED HONEY

CAPPED HONEY

If you don't have a fancy gadget that helps you test the moisture levels in your honey stores, I have a simple test you can do. Take your uncapped honey frame and tip it upside down. If liquid runs out, it's not ready. If the liquid stays in the cells, it is ready.

HONEY EXTRACTION EQUIPMENT

If you are a member of a beekeepers association, they may have equipment that members can use for free. You can also ask fellow beekeepers if anyone has equipment that you can borrow. You can buy a honey extraction machine, or just use a five-gallon bucket. If you're planning on using a five-gallon bucket, try to get one with a spout for draining the honey.

Here's your basic equipment checklist.

» Bee suit, gloves, and veil

» Smoker, lit and ready to use

» Bee brush

» Hive tool

» Uncapping tool—a little metal pick used to puncture the wax cappings on the honey cells

» Bucket, tote, or extraction machine

» Filter

» Clean bottles or jars with lids to store the honey

» Pail or pan of warm water for rinsing honey off everything

» Beekeeping journal

You can extract honey one of two ways: Either crush the honeycomb and strain it out, or spin the honey out. Either way, extracting honey will be a sticky mess. Expect this going in and set up your extraction area as if you are prepping for a bunch of toddlers to finger paint. Newspaper, plastic tablecloth, wet washcloth, cardboard, work clothes—you get the idea.

WORK AS A TEAM

After you have confirmed your honey is ready to extract, there are a few things you will need to help you. One of the main things is another person—preferably one with beekeeping experience. Beekeepers are an amazing group of people and love to help fellow beekeepers. Contact someone in your beekeepers association or a local fellow beekeeper and ask them to help you the first time you are extracting honey. You will appreciate the help of an experienced beekeeper.

HARVESTING HONEY, STEP BY STEP

Before you do anything, make sure you are prepared. Set up your honey extraction equipment far away from the beehive. Better yet, extract the honey inside your house or a building if you can. The bees won't be too happy about you taking their liquid gold, even though there is no harm in doing so. There's nothing quite like a bunch of mobbing bees to distract you from your efforts to collect their honey.

Once all of your ducks are in a row, you're ready to approach your hive. The first step in the honey harvest is to safely remove the frames from your hive.

REMOVING THE HONEY FRAMES

When removing the frames for honey, make sure you're not removing the brood frames. You don't want baby bees in your honey, nor do you want to take brood from their colony. You can tell the difference between honey stores and brood stores by the color of the caps, as described on page 90.

1. Smoke the hive as you do for inspections.
2. Remove the top board and the inner cover.
3. Inspect the frames to determine which ones have honey stores.
4. Gently brush off as many bees as possible and place the honey frames in a bucket or tote. Be careful not to crush any bees!
5. Close your hive again and take your frames to the extraction area you have set up.

OPTION 1: CRUSH AND STRAIN

Crush and strain is the slowest method of honey extraction, but requires the least amount of equipment. Pretty much anyone with a bucket and basic kitchen equipment can extract honey this way. Another bonus to this method of extraction is that it can be used with any type of beehive. With the crushing method, you take the whole comb and crush it to extract the honey, then strain it to remove all the wax.

Whether you have a bucket, tote, or some other catchment device, the goal is to remove the honeycomb from the frames and put it into your container. Many tools will work to achieve this goal, such as a sharp knife, a spoon, a spatula, or another kitchen utensil. Crush the comb to release the honey. You can use the back of a spoon, a potato masher, a spatula—pretty much any kitchen utensil that will get the job done. Then strain the wax from the honey through a fine mesh strainer. You may have to do this a couple of times to remove all the wax and particles.

The downside to crush and strain, other than the amount of time it takes, is that you destroy the comb when harvesting honey. The bees will have to rebuild their comb when you return the frames to their hive. Many beekeepers prefer this, because they like to harvest the beeswax as well as the honey.

OPTION 2: SPINNING

The spinning method uses a honey extractor. You place the frames in the extractor, then spin them using a hand crank or electricity. The centrifugal force extracts the honey within minutes.

First you will need to take your uncapping tool and uncap all the **honey cells**. It is very easy to do. Think of it like popping all the bubbles on a sheet of bubble wrap with a comb that has a lot of sharp spikes on it. All you want to do is remove or pierce the wax cap on the honeycomb so you can release the honey. When the frames are uncapped, you place them in the spinner per the manufacturer's directions, spin the frames, and pour the honey out of the spout.

The spinning extraction method is perfect for beekeepers with more than one hive or for those with more than one honey hive body that's ready to harvest. You can extract honey in minutes with the extractor, versus hours with the crush and strain method. Another bonus is that the comb is left intact. This means when you are done extracting the honey, all you have to do is return the frames to the hive. The bees will clean them up and get right back to making honey.

The downside to this method is you can only use it with Langstroth hives. Another downside is the cost of equipment. Although, as I said, if you don't have an extractor, you may be able to borrow one.

RETURNING THE FRAMES

Once you have extracted your honey, you can return the frames to the hive. There is no need to clean the extra honey off them; your bees will clean them up in no time. Follow the same steps you used to remove the frames:

honey cells cells in the comb that are used to store honey

Gear up with your safety equipment, smoke the hive, carefully remove the top, remove the inner cover, and replace the frames. Be careful not to crush any bees while you're returning the frames to the hive.

FILTERING THE HONEY

Whether you crush or spin your honey, you will want to filter out all the wax and other particles. Depending on how much wax and particles you have, you may want to strain it more than once. Once the honey is strained, you will want to keep the honey in a covered jar or bucket for a couple of days. It's important to cover your honey as soon as possible so no moisture can enter the honey and cause fermentation.

After a couple of days have passed, remove the cover from your honey and scoop off anything that has floated to the top. At this stage, your honey is ready to bottle and cap.

 ## SELLING HONEY

Each state in the United States has its own laws for selling honey. Before you sell any honey, you need to contact your local extension agency or local beekeepers association and inquire about your state's laws. Most of these laws are listed under your state cottage laws. The USDA also has specific requirements for honey labels, such as the size of the label and the mandatory weight measurements.

Honey is typically sold by weight and not by volume. Your local economy, and supply and demand, will dictate what you can charge for a pound of your honey. I generally sell my honey for $10 for 10 ounces in South Carolina. Raw honey was in such high demand in my area that I could never keep it in stock. Since you can harvest 80 pounds or more from one hive per year, you can do the math. Now you know why people are beekeepers for a living.

WHAT CAN YOU DO WITH ALL THAT BEESWAX AND HONEY?

You're about to hear me sound like Bubba from *Forrest Gump* right now when I start talking about all the amazing ways you can use honey and wax. Seriously, it's better than almost anything else. Because honey and beeswax are so versatile, even backyard beekeepers get into making products for gifts and for sale.

There are dozens more uses for both honey and beeswax than I can list here, but I'm going to mention a few to show you how amazing these products are.

HONEY

After-workout
pick-me-up

Allergy relief

As a snack with
honey straws

As a sweetener

Bake with it in
place of sugar

Coat a sore throat

Dessert topping

Dry skin relief

Eaten as it comes

Face mask
beauty treatment

Hair treatment

Hangover treatment

Treat minor burns

Treat acne

Use as a dip

BEESWAX

Artwork

Beauty products: lip
balm, lotion, body butter,
deodorant, beard wax

Candles

Cracked skin treatment

Fire starter

Furniture wax

Homemade food wrap

Leather conditioner

Preventing rust on tools

Salves

Soap

Wood conditioner

You are only limited by your imagination when it comes to uses for your honey and beeswax. Get creative and enjoy!

Chapter Nine

KEEPING YOUR BEES HEALTHY AND YOUR COLONY PRODUCTIVE

I know you got all excited about the liquid gold and maybe even the income potential from having a hive or two. However, the top priority and main focus of beekeeping is to protect the species. This means learning about the pests and diseases that can harm them. Unlike some other types of livestock, bees will generally give you a heads-up when something is wrong. I will explain what to look for during hive inspections and suggest treatment options if your bees have some health problems.

Observing your hive and noting everything in your beekeeping journal are key to helping it remain healthy. Through daily observation, you will pick up on any changes in your bees. Then, by periodically reviewing your beekeeping journal, you can compare seasons from one year to the next, or even compare hives if you have more than one.

WHY DO BEES GET SICK?

Honeybees as species have survived for thousands and thousands of years. Even though they may seem very delicate, they are actually quite resilient. Although we have done our best to domesticate honeybees, they are wild insects that are determined to survive. To this day, no one has been able to keep bees alive without allowing them access to nature.

However, over the years, increased exposure to chemicals and toxins seems to have affected their immune system. Disease treatments that beekeepers have used for the last 30 years are no longer as effective, and beekeepers are looking for alternative methods of treatment.

Many of the reasons bees get sick are completely out of human control, but some can be prevented. Some of the reasons your bees may get sick include:

» Being cooped up inside for too long

» Climate

» Exposure to bacteria and viruses

» Exposure to toxic substances

» Insects and arachnids in the hive

» Overcrowding

» Starvation

» Too much moisture

» Too much room

PREVENTING ILLNESS

As I said, some things in your environment are purely out of our control—such as the time I lost two entire colonies because our county sprayed chemicals in the area in an attempt to control mosquitoes. However, there are times when we can help prevent stress and disease. While some of these factors alone won't kill your bees, they can weaken their immune system, making them more susceptible to illness and pests.

Overcrowding can be prevented by adding boxes/**supers** to your hive, or splitting your colony to allow room for growth.

Too much room can be a problem because it makes space for predators and pests. Bees have a hard time defending and protecting a hive that is too big for them. Think of a huge Hollywood gala with lots of entry points and only one bouncer at the door. That one bouncer can't possibly keep all the riffraff out. The same principle applies when there's not enough bees and too much room in the hive. This happens when new beekeepers want to add all the boxes to their hive right away. A hive may also have too much room when it splits from a swarm, or sometimes after winter when a large portion of the colony dies. To prevent this, make sure to remove any unused boxes, and do not add a new box unless the bees have built comb on ³/₄ of the frames you already have.

super   a hive box that you add to an already established stack of hive boxes

Moisture is a problem I talked about in chapter 7, where I suggested some ways to help reduce moisture in a beehive in the winter. As a reminder:

» Don't restrict airflow into the hive, but do protect the hive from strong winter winds.

» Add an absorption blanket inside the inner cover.

» Place newspaper, cardboard, or other absorbent material in an empty box on top of the hive.

Starvation is something you can easily prevent. Your bees need water and honey, or bee food, to survive. Even when you think there may be plenty to forage on, your bees could be starving. Always have fresh water available and check their honey stores, or food supply, monthly when you perform inspections. Set up some food stores for them to overwinter as well (see page 109).

Stress on bees can come from many sources, including disease, lack of food, attacks, a new home, and too much human interference. However, one of the main sources of bee stress is lack of food and the excessive foraging that requires. This is the reason I recommended planting your bee garden *before* you set up your hives.

SIGNS OF SICKNESS

I can't stress enough the importance of observation and your beekeeping journal. Write down what you see, even if it's simple things like, "Bees are flying a lot today." You will be grateful you took good notes, trust me.

Things you can observe from outside the hive that can indicate something is awry:

» A foul smell (see AFB and EFB on page 134).

» Excessive amounts of dead bees (more than you usually see). This could mean a number of things. In this chapter, you will read about all of the pests and diseases, to rule out possible causes. Other factors could be starvation, the bees froze to death, excessive moisture, or chemical exposure.

» Lack of activity. If it's not winter and you don't notice a lot of activity, it's time to open your hive and inspect the inside.

» A lot more bee poop than usual outside of the hive.

Things you can observe from inside the hive that can indicate things aren't quite right:

» Missing queen. Contact a local or online beekeeper who sells queens, and also contact your bee mentor. A colony will not survive without a queen.

» More than one egg per cell. Too many eggs per cell mean you have workers laying eggs. Eggs from workers are unfertilized, which means they will all turn into drones. And a colony doesn't need that many drones—who will eat up all the resources without doing any of the work.

 ASK AN EXPERT

I mentioned at the very beginning of this book that I recommend attending beekeeping classes with your local beekeepers association. One of the benefits of attending a class like this, other than education, is finding a mentor—or at the very least, meeting beekeepers with years of experience who are happy to share their knowledge.

Mentors volunteer to work with new beekeepers to help them succeed. Beekeepers are a different breed. It is the one true

hobby/profession where I never witnessed jealousy or people hoarding information for themselves. They all have the exact same goal: to help the bees. So by helping new beekeepers, they are helping even more bees than they can raise themselves. If you don't have a beekeepers association, try to contact a local beekeeper and ask them if you can reach out to them if you ever have any questions. I'm betting they will be happy to help.

You should contact your mentor or another experienced beekeeper any time you feel something is off. My mentor was my saving grace. I called or texted him any time I had a question, and he was happy to help. Of course, you will want to use common sense and not call them every day. Never feel any question is dumb or too embarrassing to ask. In a couple of years, you will be that mentor to a new beekeeper and know exactly where they are coming from.

» Foul smell. The foul smell coming from a diseased brood is different than the stinky smell I was referring to during the fall honey harvest. A foul smell could mean you have AFB (see page 134).

» Funny-looking brood, meaning anything that looks different than normal.

» The bees aren't cleaning their hive.

» The bees are forming supersedure cells or multiple queen cells, or are creating queen cells out of ordinary brood cells.

BEES AND NUMBER TWO

I'm willing to bet you've never given any thought to where bees poop. Am I right? For the most part, bees do their business outdoors. These trips to the potty are called **cleansing flights**. Remember those little sprinkles of "rain" I was talking about earlier? Well, now you also have bee poop to contend with. I swear, it's not safe to look up in the sky anymore.

All kidding aside, you can tell a lot about your colony's health by their poop. Bee poop will look a lot like tiny brown skid marks on the outside of their hive. Generally, they will fly far enough away from their hive to do their business. However, when ya gotta go, ya gotta go. If your bees have excessive little brown marks on the outside of their hive, they could have diarrhea or dysentery.

Dysentery in bees can be caused by pathogens like nosema disease (a fungal disease that mainly affects bees), or too much solid-waste buildup. Nosema by itself does not contribute to a bee's death, but dysentery can. Bees are able to store 30 percent to 40 percent of their body weight in waste. This helps them survive the cold months when the temperature is too low for their cleansing flights. If a bee consumes too many solids from its honey, it can get constipated. When the bees come out

cleansing flight a quick flight outside the hive to poop from hibernation in the spring and consume a lot of water, things start to flow. You may notice more bee poop and dead bees during their spring cleaning. Just keep a watchful eye for what looks like more than usual.

THE MOST COMMON PESTS AND WHAT TO DO ABOUT THEM

Pathogenic mites, fungi, bacteria, viruses, and pests all can plague beehives. Despite all of the turmoil that can attack bees, a healthy hive can do a phenomenal job of protecting its colony. Hives have cleaners who work nonstop at cleaning the hive and all the bees. They also have guard bees to protect the entrance to their home.

There will be times, though, where you need to step in and give them some help. If you run into a situation where you must treat your hive, make sure to rotate your treatments so the pests and bacteria don't build up resistance to whatever you are using.

AMERICAN FOULBROOD (AFB) AND EUROPEAN FOULBROOD (EFB)

AFB and EFB are highly contagious and infectious diseases caused by spore-forming bacteria. They do not affect adult bees, but affect the larvae and pupae. This disease gives the hive a foul odor, hence the name.

In addition to the odor, you will see the evidence in the brood cells. Normal larvae are pearly white, but larvae infected with AFB turn a caramel-chocolate color and melt into a gooey mess on the floor of the cell. They invariably die after the cell is capped. EFB is less severe than AFB (although still very dangerous). EFB-infected larvae will be off-white, progressing to brown, and are twisted inside the brood cell. They die before the cell is capped.

ALSO LOOK FOR:

» A spotted pattern on the brood

» Sunken cappings on the brood cells

» Off-center holes in the brood cell caps

» Scales on the larvae

» Caramel color of dead larvae

If you suspect you have AFB or EFB, you must contact your local extension agency or local beekeepers association. Handling and disposal of infected hives may be regulated by state laws. Unfortunately, if your colony gets AFB or EFB, there is no saving it. Once you've contacted your county extension, either they will come to dispose of your hive or they will instruct you on how to do so. Extreme measures are required because the spores are highly contagious and can remain viable indefinitely on beekeeping equipment.

VARROA MITE

The Varroa mite is a parasite that sucks the blood of honeybees and brood. It can only reproduce in a honeybee colony. These vampire arachnids weaken the bees and

VARROA MITE

shorten their life. The adult female mite has a reddish brown body shaped like a flattened oval, and eight legs. You should be able to see these pests with the naked eye.

There are many chemical treatments you can use for Varroa mites. They include Apivar, Hopguard II, CheckMite+, Apistan, Formic Pro, Mite Away Quick Strips, Api Life Var, and oxalic acid. Each treatment will come with its own application instructions.

DEFORMED WING VIRUS

This is a viral disease associated with Varroa mite infestations, although it can also be seen in colonies that haven't been affected with Varroa. It may cause bees to form distorted, misshapen, twisted, or wrinkled wings. Obviously, these bees can't fly and can't help their hive. The most effective way to deal with deformed wing virus is to control the Varroa mite population.

SACBROOD

Sacbrood is a viral disease that affects a honeybee brood, mainly worker bee larvae. Sacbrood virus causes an uneven brood pattern, with irregular cappings found throughout the brood cells. This disease is relatively easy to recognize because it prevents the brood from developing to the pupa stage.

Generally, a strong hive can overcome sacbrood disease without the beekeeper's interference, although severe cases may require replacing the queen.

BEE LOUSE

The bee louse is a small, wingless member of the fly family that is occasionally found on bees. It is not a bee disease, but does subject colonies to a lot of stress, thus weakening it.

The bee louse adults look kind of like Varroa mites, but they have only six legs. To treat for adults, add a little bit of tobacco to the smoker and smoke the hive. The larvae look like little worms, but they are very hard to see with the naked eye. You are more likely to see comb damage than the actual larvae. To treat for bee louse larvae, you will want to wrap the frames in plastic and freeze for 48 hours. This will ensure you kill the adults and the larvae.

TRACHEAL MITE

This tiny arachnid is a parasite of the honeybee that infects the respiratory system. It's capable of infecting drones, worker bees, and queen bees. It reproduces inside the bee's breathing tubes and feeds on the bee's blood, killing it. Some signs are increased winter deaths and lack of spring brood production. Your county extension office may offer a test if you suspect you have tracheal mites.

Suggested treatment is a multistep process of using menthol and grease patties made of vegetable shortening and sugar (which interfere with mite reproduction), in addition to **requeening** the hive.

requeening deposing the old queen and replacing her with a new one

WAX MOTH

Wax moths are a pest that damages beeswax combs, comb honey, and bee-collected pollen. They're mainly an issue for weak colonies. Wax moths are something that all hives will deal with at one point or another, even if you're the best beekeeper in the world.

To help prevent wax moths from taking over your hive, reduce the amount of unused space in your hive. Don't add boxes unless your hive needs them. If no one is home to defend the hive, wax moths will move in and take over. Another way to treat wax moths is to wrap the infested frame in a plastic bag and freeze it for 48 hours. You can

WAX MOTH INFESTATION

also fumigate your hive with paradichlorobenzene (PDB) crystals.

SMALL HIVE BEETLE

This invasive pest inhabits nearly all honeybee colonies. Given the chance, small hive beetles will multiply and cause serious damage to your colony. Like the wax moth, almost every beekeeper will have to deal with them.

There are some chemical treatments, but prevention is the best defense against small hive beetles. As with the wax moth, limit the amount of available room in the hive. If you are using a Warre or Langstroth hive, add a board with a wire bottom and a tray below it to the bottom of your hive. Put some dish soap on the tray. The beetles will fall in the tray and will be unable to get out.

CRITTERS TO WATCH OUT FOR

Are you beginning to wonder how bees survived all these years? My guess is that they are really good at survival. I would dig myself in a hole and never come out if I had this many creatures after me. It's that liquid gold that no other insect or animal can make that drives critters of all shapes and sizes crazy, and turns them into hive raiders. While you can't guarantee their protection, there are some things you can do to protect your bees.

» **Ants**: Sprinkle cinnamon around the base of the beehive. I have used this method and it works great—although you do have to repeat it often. We buy several large containers of cinnamon from the dollar store to keep plenty on hand.

» **Other bees**: See Robber Bees below.

» **Spiders**: It is said that spiders don't like mint. You can plant some around your hive to repel spiders and other insects. Make sure to remove any spiderwebs on or around your beehives.

» **Mice**: If it's not hot outside, you can use the entrance reducer if you're having trouble with mice getting in your hive. Mice won't hurt the bees, but they will eat their comb.

» **Frogs**: You can try to move a frog's habitat, such as damp areas, rocks, and things they can hide under,

 ## ROBBER BEES

You may find this a bit surprising, but bees will rob other bees of their honey. Even in the insect world, they want what their neighbor has. Although bees are extremely dedicated to ensuring the colony survives, they only care about their own colony's survival. A stronger bee colony will rob a weaker colony of its honey, especially during the fall or other seasons when the nectar flow is low.

As a new beekeeper, it may be hard to tell if another hive is robbing your hive. However, there are signs of this thievery. For instance, if you see wrestling at the front of the hive, or two bees attached together at the front of the hive, this is a sign of robbers. Another indication is seeing dead bees with stingers by the front of the hive in the fall. (Dead bees by the front of the hive without stingers in the fall are the drones

getting kicked out before winter.) If you noticed that the honey stores have chew marks on the comb and the cap has been punctured or removed, this is also an indication of thieves.

If you notice robbing, you can help your hive by installing the entrance reducer if it's not hot, or small mesh wire across the front if it is hot. This will give the security bees less space they have to defend.

farther away from your beehives. In all honesty, frogs won't eat enough bees to hurt the colony—unless they eat the queen, of course.

» **Birds**: Birds will eat bees in flight, but they will leave your hive alone. Like the frogs, they generally won't eat enough to harm your colony.

» **Other mammals**: To keep away raccoons, weasels, badgers, skunks, foxes, and maybe bears, you can attach flooring tack strips, like the kind used to hold the carpet down, to plywood. Place these strips down in front of the hive, so anything walking up to the hive will step on the tacks.

Chapter Ten

THE FUTURE OF YOUR BEES

As you work through your first year of beekeeping, you'll transition from greenhorn to an experienced beekeeper. Each season you'll learn new things; even beekeepers with 20 years of experience are constantly learning something new. That is the amazing thing about beekeeping: I've never met a beekeeper who got bored or tired of it.

Don't get me wrong, it's not always easy. When we lost two colonies to chemical drift, I wanted to throw in the towel. We cleaned our hives and put them behind the shed, never to be used again. However, nature had other plans; nature wanted us to be beekeepers. Months after we put our hives in storage, we walked behind the shed and heard a muffled buzzing sound. Lo and behold, feral bees had taken residence in our hives! Free, strong, glorious honeybees. My heart was full again. Not to mention, feral bees often form the strongest and most resilient colonies.

DO YOU NEED MORE THAN ONE HIVE?

One thing I didn't touch on earlier was how many hives you should have. Remember, this book wasn't written with the commercial beekeeper in mind—I'm talking to you, backyard beekeepers. I am going to suggest (strongly) that you get at least two hives. Why two? Everything is better in pairs. Even beehives.

» **Compare and contrast:** If you're a first-time beekeeper, you will benefit by having something to compare to. If you have one hive that looks off to you, you can compare it to the other to see if that's "normal."

» **You have a backup:** If you lose a hive or a queen, for whatever reason, you will have a backup. I have lost one hive in the past and the remaining hive split with a swarm in the spring, which gave me two hives again. If you lose one and that was your only hive, you will have to start all over.

» **Combine efforts**: If you have a weak hive and have only one, you will have to requeen your hive. Then there is still a chance it may not survive. However, if you have two hives and one is weaker, you can combine the hives to make one strong hive.

» **Save money**: Believe it or not, twice the number of hives does not cost twice the amount of money. For instance, you still only need one bee suit, one hive tool, one honey extractor, and so on. You can share many of the expenses between the two hives.

» **Double the honey**: Two hives mean double the liquid gold, which also means double the money, if you plan on selling your extras.

DON'T FEAR THE SWARM

Swarms can be a welcome or a feared sign by beekeepers. Those who wish to expand their apiary are excited to see the signs of a swarm, because this means they get another bee colony for free. Just to review, a swarm is when half your colony and the old queen leaves the hive in search of a new home. The remainder of your bees and the new queen will work on building their colony. When a colony begins to swarm, it means they are healthy and growing, so that is a good sign. However, if you don't want your colony to grow, you may be sad to see them go.

Watching a swarm is truly an amazing sight. When our bees swarmed, we could hear them all the way inside the house. Their collective noise is very recognizable once you know what it sounds like.

The swarm will contain the original queen and about half the workers. They will start to hang out on the outside of the hive in a cluster, which is different than bearding. Once they leave the hive they will remain in a cluster up high somewhere until the scouts find them a new home. During a swarm, the bees are very docile; they are focused on protecting the queen and have little use for you. You are least likely to get stung by bees when they are swarming.

In addition to the obvious external signs of swarming, another sign is increased queen cells in the hive. If you don't want your bees to swarm, one way to help prevent it is to provide them with extra room. Add frames to your hive and give them room to grow.

If you see signs of swarming, contact your beekeeping mentor or local association. They can help you capture the swarm for yourself, or you can give the swarm to another beekeeper. It's a great way to multiply the bee (and honey) population.

A ROYAL COUP D'ÉTAT

Your queen is *the most important* bee in the colony. Without her, the colony would fail. The queen determines the health, the growth, and the survival of the colony. Of course, every bee has an important and vital role, but none as great as the queen's. Because she is so important, there may be times when you need to depose the queen and/or put a new queen on the throne—a process beekeepers call requeening.

» **Queenless:** You've inspected your hive and noticed it doesn't have a queen. Maybe you can't find her, or maybe you see other signs of a queenless hive. Multiple eggs in one cell is a sign of a queenless hive; it means the worker bees are now laying eggs. Multiple queen cells forming can also be another sign of a queenless hive.

» **Squished the queen:** It's unfortunate, but it does happen. Maybe it happened during a hive inspection or extracting honey or even when you were replacing the frames. Accidents happen, and when they do, you need a new queen ASAP.

» **Aging queen**: Although queen bees can live three to four years, they are the most productive in years one and two. To maintain a strong and productive hive, I suggest replacing your queen after her second year, if the hive hasn't already done so on its own.

» **Rejection**: It is rare, but there are cases where the colony simply rejects the queen.

» **Sick queen**: The queen is fed by the worker bees. In some cases, the workers transfer sickness or disease to the queen.

» **Lack of pheromones**: The workers in the hive react to a queen's pheromones. If the queen fails to produce enough pheromones, the hive will go all willy-nilly.

REQUEENING

If your queen is still in the hive and you need to replace her for one reason or another, it's important to have the new queen in hand before getting rid of the old one. You can order a queen through a supplier or contact a beekeepers association to see if anyone raises queens for sale. From personal experience, new queens generally cost around $30 or more.

If you think about it, the queen is related to everyone in the colony. Her DNA runs through the veins of every bee. They have a personal connection. When the queen needs to be replaced for any reason, the hive must go through the acceptance process all over again, just as if you were installing a new pack of bees with a caged queen. I've described that process in detail in chapter 5.

When you've installed the new queen into your hive, you will need to dispose of the old queen. Off with her head! I know it sounds harsh, but it's necessary sometimes for the health and survival of the colony.

KEEP THINGS CLEAN

The final bit of knowledge I'm going to pass on to you is about keeping everything clean. Do you know the saying "Cleanliness is next to godliness"? It's certainly true of a beehive. Cleanliness is extremely important to a healthy beehive. *Extremely* important. In fact, many of the diseases that kill beehives are transferred from the beekeeper to the beehive.

Pretty much every piece of bee equipment you have must be cleaned regularly. You can clean most of the equipment with hot water and soap. I like to use more natural cleaners like washing soda, vinegar, and baking soda. I use bleach only when necessary.

» **Hive tools:** To clean your hive tools, mix vinegar or bleach (not both) with water and put it in a spray bottle or a bucket. You can spray your tools and wipe them clean, or soak them and dry them off. This includes your bee brush.

» **Bee feeder:** Wash in warm, soapy water and make sure to rinse well. Do this every time you add new bee food. If your colony was sick, use a bleach solution to clean your feeder.

» **Water dishes:** Wash in warm, soapy water and make sure to rinse well. Do this at least once a week. If your colony was sick, use a bleach solution to clean your feeder.

» **Bee suit, gloves, and veil:** Your bee suit and gloves should come with a label with wash instructions. For the veil, I always handwash it in the sink with warm, soapy water, then I rinse and line dry it.

» **Beehive and frames:** If you are removing a frame or box, replacing a board or extracting honey, you will need to wash your hive—especially if you lost a hive due to sickness or disease. If you bought a used hive or someone gave you a hive, it is extremely important you wash it before adding bees. Add one cup of bleach per five gallons of warm, hot water to a large bucket or tub. Use a wire scrub brush to remove any debris. Soak everything in the bleach water for 15 minutes. Remove, scrub again if needed, and repeat the soaking. When everything is clean, you can place it in a sealed plastic bag for storage or let it dry completely and place it back in the hive.

» **Honey extraction equipment**: Clean all of the honey extraction equipment with warm, soapy water or warm water mixed with washing soda. Make sure to get all of the honey off your equipment. Rinse thoroughly and dry before returning to storage.

» **Smoker**: Keeping your smoker clean and free of buildup is just as important as keeping your fireplace chimney clean. With a wire brush, scrape the inside walls of your smoker. Remove any loose debris. Use your vinegar spray, spray the rest of the inside of your smoker and sprinkle with baking soda. Rinse and repeat. Be careful not to get the pump part of your smoker wet. I have found that small wire brushes about the size of toothbrushes work the best for this. Clean your smoker after a couple of uses or as soon as you notice any black buildup on the inside.

Sanitary beekeeping practices will help protect your hive from spreading disease and sickness. However, if you are dealing with AFB or EFB, don't try to clean this yourself. Contact your county extension. They may require you to burn your hive or frames, and they will tell you how to sanitize all your equipment.

ARE YOU READY TO START BEEKEEPING?

So here we are at the end. I've shared what I know from my years of experience, and now you have all the knowledge you will need to get started and be a successful beekeeper. I'm genuinely thrilled that you are taking these steps to help preserve insects that are vital to our survival. It's mind-blowing to think of it in that regard, but it's true. We need bees, and bees need us.

Whether you want to be a backyard hobbyist or a commercial beekeeper, every person can make a difference. You will soon learn that everyone you meet will be excited that you are keeping bees. The honey requests will start rolling in even before you have your beehive set up.

Remember to learn from others, lean on your local beekeepers for support, and come back to this book to troubleshoot and help you with any problems you may face. The only thing I can't provide you with is actual hands-on experience. But you'll get there.

BEEKEEPING RESOURCES

ORGANIZATIONS

**How to Find Your County
   Extension Office**
Look up your local county extension
office by state.

To find your local county extension
agent's office, just go to the website and
click on your state.

PickYourOwn.org/CountyExtension
 AgentOffices.htm

American Beekeeping Federation
For more than 75 years, the American
Beekeeping Federation (ABF) has been
working in the interest of all beekeepers,
large and small, and those associated
with the beekeeping industry to ensure
the future of the honeybee.

ABFNet.org

American Bee Journal
Find your local beekeepers association
by state.

AmericanBeeJournal.com/tiposlinks
/beekeeping-associations

BEEKEEPING APPS

Apps change all the time, but some you
   may want to try:

» American Bee Journal

» Apiary Book

» Beekeeping Pro

» HiveTracks

» BeePlus Beekeeping Manager

» BeeRM

» Hive Tool Mobile

» HiveKeepers for Beekeepers

BEEKEEPING SUPPLIES

Dadant and Sons
Dadant.com

Kelley Beekeeping
KelleyBees.com

Mann Lake
MannLakeLTD.com

INFORMATION

Ag Daily
www.AgDaily.com/crops/are-honey
-bees-endangered

**American Association for the
Advancement of Science**
AAAS.org/news/science-begins
-unlock-secrets-brains-age-well

**Cornell University Department of
Entomology**
Entomology.CALS.Cornell.edu
/extension/wild-pollinators
/native-bees-your-backyard

Harvard University Press
www.HUP.harvard.edu/catalog
.php?isbn=9780674418776

Honey Bee
HoneyBee.DrawWing.org

Kentucky Farm Bureau
KYFB.com/federation/newsroom
/honey-bees-livestock-with-wings

Lund University
https://phys.org/news/2017-10
-bees-home.html

Michigan State University
https://Pollinators.MSU.edu/resources
/beekeepers/diagnosing-and-treating
-american-foulbrood-in-honey-bee
-colonies

Natural Resources Defense Council
NRDC.org/sites/default/files/bees.pdf

New Phytologist Trust
NPH.OnlineLibrary.Wiley.com/doi
/full/10.1111/j.1469-8137.2009.02925.x

**Northwest Jersey Beekeepers
Association**
NJBeekeepers.org/nwba/documents
/Harvesting%20Honey%20Notes.pdf

The Pennsylvania State University
https://extension.psu.edu/a-quick
-reference-guide-to-honey-bee
-parasites-pests-predators-and
-diseases

Scientific Beekeeping
ScientificBeekeeping.com
/sick-bees-part-1

Tennessee Department of Agriculture
TN.gov/agriculture/businesses/bees
/apiary-registration.html

University of Arkansas Division of Agriculture
https://www.uaex.edu/farm-ranch
/special-programs/beekeeping
/default.aspx

University of California
UCANR.edu/blogs/blogcore/postdetail
.cfm?postnum=14566

University of Georgia
http://bees.caes.uga.edu/bees
-beekeeping-pollination/honey
-bee-disorders/honey-bee
-disorders-bacterial-diseases.html

US National Library of Medicine
NCBI.NLM.NIH.gov/pmc/articles
/PMC5681286

GLOSSARY

American foulbrood (AFB) and European foulbrood (EFB) AFB and EFB are highly contagious and infectious diseases caused by a spore-forming bacterium; they are fatal to your bee brood

apiary a place where honeybees are kept

attendant bees worker bees that take care of the queen; they groom her and feed her

bearding when a large number of bees congregate on the front of the beehive, making it look like the hive has a beard; they do this to cool the interior of the hive

beauticians worker bees that clean off debris and groom the other bees

beebread a mixture of collected pollen, bee secretions, and nectar or honey that is a food source for bees, especially for eggs and larvae

bee brush a long brush used for gently brushing bees into or off the hive and frames

bee eggs fertilized eggs laid by the queen that will develop into bees

bee food a sugar water supplement used to feed your bees when nectar is not available

bee louse a small, wingless member of the fly family that is occasionally found on bees

bee space the ideal amount of space (3/8 inch) between structures in a beehive

bee suit a protective suit beekeepers wear to avoid getting stung; generally they're white, which is a calming color to bees

beekeeping journal a book used for all of your beekeeping records

beeswax a natural wax produced by bees to make their comb

brood the developing bee cells; they might be eggs, larvae, or pupae, and drones, workers, or queens

caged queen a queen that is not the colony's queen, and can be bought seperately; she comes in a cage and must be freed by the worker bees

candy plug a sugary plug sealing the bottom of a caged queen's cage

capped cells cells that are filled with either brood or honey

cells honeycomb consists of hexagonal walls of wax, each enclosing a cell

cleansing flight a quick flight outside the hive to poop

cluster when bees huddle in the center of a hive to keep warm

colony a collective group of bees that live together; it consists of one queen, drones (male bees), workers (female bees), eggs, larvae, and pupae (developing bees)

colony collapse disorder an unexplained occurrence when either all or most of the bees in the hive disappear

comb a group of six-sided cells made of beeswax where bees store their honey and pollen and raise baby bees; the comb is made up of two layers that are attached at the base

construction crew the worker bees that build the honeycomb and repair damaged combs

deformed wing virus a viral disease associated with Varroa mite infestations

drawn combs cells that have been built out by honeybees from a foundation in a frame

drawn frame (or drawn comb) a frame on which the bees have built the comb from beeswax; it is now ready for pollen, honey, or brood

drone a male bee; its only job is to mate with the queen

entrance reducer a small block of wood that reduces the size of the entrance to the beehive

established colony also known as a full hive, this consists of a fully established colony with drawn combs on frames, with wax, brood, pollen, honey, drones, workers, and a queen bee that is the queen of that colony

extractor a machine used to extract honey by spinning the frames

fixed-comb hive a beehive without removable frames

foragers worker bees that gather nectar and pollen

foundations the wax forms installed in the frames of hives; the bees then build their comb off of these foundations

frame racks or frames inside a hive on which the bees make honeycomb

full hive also known as an established colony

full super eight to ten established drawn combs on frames, with wax, brood, pollen, honey, and bees

hive beehives are structures made by people to house bees; technically, humans make a hive, while bees make a nest

hive body the main box part of the hive, where the bees live

hive tool a tool that resembles a pry bar and is great for prying open the hive box, scraping propolis, moving frames, and more

honey a sweet food substance produced by bees

honey cells cells in the comb that are used to store honey

honey flow the time when bees produce a large amount of honey

honey sac a gland at the end of a bee's esophagus in which nectar gathered while foraging is stored

honeybee a flying insect that produces wax and honey

house bees worker bees inside the hive that gather the nectar and pollen from the foragers

housekeepers worker bees that perform every duty a housekeeper would: take out the trash, remove dead bugs, clean the honeycomb, and keep things nice and tidy

inspection when you open a beehive to inspect the inside and see what your bees are up to

larva (plural, larvae) a white, legless, grublike insect that represents the second stage of bee metamorphosis

livestock food-producing animals; the honeybee is classified as livestock

moisture quilt also called an absorption blanket, it's something made of absorbent material and added in cold weather to the top of the hive, where condensation tends to collect

morticians worker bees that remove the dead bees from the hive

nectar a sweet liquid produced by flowers that bees collect to make honey

nectar collectors worker bees that collect nectar for the hive

nectar flow the time of the year when flowers produce nectar

nuc an already established frame with wax, brood, pollen, honey, and bees; nucs include drones, workers, and a queen bee

nurses worker bees that feed and care for the growing larvae

orientation flights when new bees take their first flights out of their hive to get to know their surroundings

pack an actual pack of bees, generally weighing around three pounds; it will contain both worker and drone bees, and should include a caged queen

pheromones bee pheromones are chemical messengers produced by bees that they use to communicate with one another

pollen a yellow to orange powdery substance produced by plants to facilitate pollination

pollen basket an area on the hind legs of honeybees where they collect pollen while foraging to transport it back to the hive;

hairs hold the nectar-moistened pollen in place

pollination the act of fertilizing plants by transferring pollen from a male plant to a female plant

pollinators animals or insects that pollinate plants by transferring pollen from one plant to another

propolis also known as bee glue, it's a substance bees make from saliva and botanical sources to protect the hive

pupa (plural, pupae) the third stage in honeybee metamorphosis, during which it changes (pupates) from a larva to an adult bee

queen an adult female bee that is the only one in a hive capable of laying fertilized eggs

queen cells cells that are formed when the workers need to replace the queen; they are larger than the other cells

queenless a beehive without a queen; the bees will not be able to reproduce, and the hive is doomed

queenright a colony that has a queen that is producing eggs and is productive

quilting box a box added to the top of the hive that contains a filler that helps absorb moisture from condensation during the colder months, keeping the bees drier and warmer

raw honey honey that has not been heated and pasteurized as part of commercial production

refractometer a tool used to measure the moisture level in a hive or in the honey

removable-frame hives hives with frames that can be taken out, allowing beekeepers full access to their hives for inspection and honey collection

requeening deposing the old queen and replacing her with a new one

royal jelly a nutritious secretion produced by the nurse bees and fed to all of the larvae

sacbrood a viral disease that affects honeybee brood

scouts worker bees that search for a new source of pollen, nectar, propolis, water, or a new home

security worker bees that protect the hive from unwanted guests

small hive beetle an invasive pest that inhabits nearly all honeybee colonies

smoker a small device that produces smoke; used to keep bees calm

super a hive box that you add to an already established stack of hive boxes

supersedure cell a cell for a queen bee that is made when the existing queen needs to be replaced for health reasons

swarm when half of a bee colony leaves with the old queen to form a new colony

swarm cell   when the colony has expanded and they are ready to swarm to make room in their hive, the workers will develop swarm cells; when the new queen emerges, a portion of the colony will leave with the old queen

tracheal mite   a parasite of the honeybee that infects the respiratory system

uncapping   when you remove or puncture the beeswax cap of sealed comb cells

uncapping tool   a little metal pick used to puncture the wax cappings on the honey cells

Varroa mite   a parasite that attacks honeybees and brood

veil   a hat with netting, used by beekeepers for protection against stings to the face, head, and neck

waggle dance   a series of movements that bees use to communicate with each other; the duration and direction of the bee dance can instruct other worker bees where to locate food

water girls   worker bees that collect water for the hive

wax moth   a pest that damages beeswax combs, comb honey, and bee-collected pollen

worker bees   female bees; they do all the work in the hive: gathering nectar, producing honey, guarding the hive, caring for the queen and the larvae, and keeping the hive clean

REFERENCES

CHAPTER 1

Hageman, Markie. "Are honey bees endangered? Here's the truth of the matter." AgDaily. August 1, 2018. https://www.agdaily.com/crops/are-honey-bees-endangered

Kentucky Farm Bureau. "Honey Bees: Livestock with Wings." December 5, 2017. https://www.kyfb.com/federation/newsroom/honey-bees-livestock-with-wings

Natural Resources Defense Council. "Why We Need Bees." Bee Facts. March 2011. https://www.nrdc.org/sites/default/files/bees.pdf

O'Neil, Kathleen. "Science Begins to Unlock the Secrets of Brains that Age Well." American Association for the Advancement of Science. February 17, 2018. https://www.aaas.org/news/science-begins-unlock-secrets-brains-age-well

UC Irvine Institute for Memory Impairments and Neurological Disorders. "The 90+ Study." UCI Mind. 2019. http://www.mind.uci.edu/research-studies/90plus-study

University of California, Santa Barbara. "Why is it that Honey will never spoil or 'go bad'?" UCSB ScienceLine. June 2, 2010. http://scienceline.ucsb.edu/getkey.php?key=2379

Wikipedia. "Pollen." Updated February 18, 2019. https://en.wikipedia.org/wiki/Pollen

CHAPTER 2

Betterbee. "Glossary of Beekeeping Terms." 2019. https://www.betterbee.com/glossary

Cornell College of Agriculture and Life Sciences. "Ground nesting bees in your backyard!" Department of Entomology. 2019. https://entomology.cals.cornell.edu/extension/wild-pollinators/native-bees-your-backyard

Gempe, Tanya, PhD, and Martin Beye, PhD. "Sex Determination in Honeybees." *Nature Education* 2, no. 1

(2009). https://www.nature.com /scitable/topicpage/sex-determination -in-honeybees-2591764

James, Brad. "Pollen Grains vs. Bee Bread (Fermented Bee Pollen): What's the Difference?" Beepods. 2018. https:// www.beepods.com/bee-pollen-vs -bee-bread

Pawelek, Jaime. "Urban Bee Legends." UC Berkeley Urban Bee Lab. Accessed February 28, 2019. http://www .helpabee.org/urban-bee-legends.html

Price, Robbie l'Anson, and Christoph Grüter. "Why, when and where did honey bee dance communication evolve?" *Frontiers in Ecology and Evolution* (November 2015). https:// www.frontiersin.org/articles/10.3389 /fevo.2015.00125/full

von Frisch, Karl. *The Dance Language and Orientation of Bees.* Leigh Chadwick, trans. London: Belknap Press, 1967.

CHAPTER 3

Bartel Honey Farms Inc. "The Amazing Honey Bee." Accessed February 28, 2019. http://www.bartelhoneyfarms.ca /index/Bee_FAQs_(and_Facts).html

Betterbee. "Bees and Nucs." 2019. https:// www.betterbee.com

Dadant. "Catalog." 2018. https://www .dadant.com

Lund University. "How Bees Find Their Way Home." October 17, 2017. https:// phys.org/news/2017-10-bees-home .html

Mann Lake. "We Know Bees." 2018. https://www.mannlakeltd.com

Skinner, John. "What is 'Bee Space' and why is this important to beekeeping?" Extension. November 10, 2009. https:// articles.extension.org/pages/44117 /what-is-bee-space-and-why-is -this-important-to-beekeeping

Thoo, Mandy. "Scientific breakthrough simply un-bee-lievable." Science Alert. June 18, 2012. https://www.sciencealert .com/scientific-breakthrough-simply -un-bee-lievable

CHAPTER 4

Dadant. "Purchasing Established Hives." 2018. https://www.dadant.com/learn /purchasing-established-hives

James, Brad. "Honey Bee Breeds and Their Attributes." Beepods. 2018. https:// www.beepods.com/types-honey-bee -breeds-attributes

CHAPTER 5

Beesource. "Theory behind lemon grass oil." Accessed February 28, 2019. https://www.beesource.com/forums /showthread.php?320728-Theory -behind-Lemon-Grass-oil

Da Silva, Cristiane de Bona, Sílvia S. Guterres, Vanessa Weisheimer, and Elfrides E. S. Schapoval. "Antifungal activity of the lemongrass oil and citral against *Candida* spp." *Brazilian Journal of Infectious Diseases* 12, no. 1 (February 2008). http://www.scielo.br /scielo.php?script=sci_arttext&pid =S1413-86702008000100014

Tennessee Department of Agriculture. "Apiary Registration." Accessed February 28, 2019. https://www.tn.gov /agriculture/businesses/bees/apiary -registration.html

Tofilski, Adam. "Nasonov gland." Honey bee. 2012. http://honeybee.drawwing .org/book/nasonov-gland

CHAPTER 6

Plonski, Sarah. "Beekeeping 101: Learn How to Use the Bee Smoker." The Honeybee Conservancy. Accessed February 28, 2019. https://thehoneybee conservancy.org/2017/09/03/bee -smoker

CHAPTER 7

Mao, Yun-Yun, and Shuang-Quan Huang. "Pollen resistance to water in 80 angiosperm species: Flower structures protect rain susceptible pollen." *New Phytologist*, July 17, 2009. https://nph .onlinelibrary.wiley.com/doi/full/10.1111 /j.1469-8137.2009.02925.x

CHAPTER 9

University of Georgia Honey Bee Program. "Bacterial Diseases." Bees, Beekeeing & Pollination. 2018. http://bees.caes.uga .edu/bees-beekeeping-pollination /honey-bee-disorders/honey-bee -disorders-bacterial-diseases.html

INDEX

A

American foulbrood (AFB), 134–135, 148
Apiaries, 3, 42

B

Backyard beekeepers, 33
Bearding, 86, 105–106
Beebread, 25
Bee breeders, 33
Bee brushes, 47, 49–50
Bee feeders, 51, 78–79, 147
Bee food, 69, 78, 109
Beehives. *See* Hives
Bee inspectors, 33
Beekeeping
    benefits, 3–6
    challenges, 7–9
    classes, 132
    costs, 8
    getting started, 13
    legal considerations, 9–10, 42, 67, 125
    other considerations, 10–13
    supplies, 46–51
    types of beekeepers, 32–34
Bee louses, 136–137
Bees. *See also* Sickness, in bees
    buying, 55–57, 59, 65–66

drones, 3, 20–22, 25–26, 107–108
feeding, 76, 78–79
free-bees, 57–58
gender determination, 19
life cycle, 24–26
queens, 12, 15–17, 26, 72–73, 83, 137, 145–146
races, 53–55
scouts, 22
in the wild, 22–23
workers, 6, 17–20, 25
Bee space, 36
Bee suits, 9, 48, 49, 71, 147
Beeswax, 126–127
Broods, 24, 90

C

Caged queens, 72
Capped cells, 25
Carniolan honeybees, 54–55
Cells, 24–26, 90–91, 124
Children, 12, 31
Cleanliness, importance of, 147–148
Cleansing flights, 133
Clustering, 84, 104
Colonies, 3, 57, 75–76
Colony collapse disorder, 15
Combs, 23, 90–91
Commercial beekeepers, 33
Communication, 6, 23

Cross-pollination, 6
Crush and strain extraction method, 122

D

Dead bees, 91, 103–104, 131–132, 139
Deformed wing virus, 136
Drones, 3, 20–22, 25–26, 107–108
Dysentery, 133

E

Eggs, 90, 132
Entrance reducers, 70, 105
Environmentalist beekeepers, 32
Equipment, 46–51, 120–121, 147–148
Established colonies, 57, 75–76
European foulbrood (EFB), 134–135, 148
Extraction methods
    crush and strain, 122
    spinning, 124
Extractors, 50

F

Fall, 107–109
First aid, 97–99

Foundations, 37
Frame grips, 51
Frames, 24, 147–148
Frisch, Karl von, 23

G

Gardens, 59–61
Gloves, 48, 49, 147

H

Harvesting honey
  best time of year for, 118
  extraction equipment,
    120–121
  knowing when it's time to,
    118–119
  step-by-step, 121–125
  uncapped honey, 119–120
Health benefits, 5–6
Hive body, 37
Hives. *See also* Inspections;
  Observations
  benefits of multiple,
    143–144
  buying, 43–45
  cleaning, 147–148
  deciding on type, 39
  defined, 5
  DIY, 42, 45–46
  established colony instal-
    lation, 75–76
  fixed-comb, 34–35, 42
  Langstroth, 35–37, 43, 70,
    93–94
  location, 30–31
  nuc installation, 74–75
  pack installation, 73
  registering, 67
  releasing queens, 71–73,
    83
  removable-frame, 34
  reorienting, 76
  roofs, 29

size, 30
top bar, 37–40, 43, 94
transferring bees to,
  69–71
Warre, 41–43, 95
Hive tools, 47, 147
HiveTracks beekeeping app,
  97
Honey
  fall, 108–109
  flavoring, 60
  harvesting, 117–125
  -hijackers, 23
  how it's made, 115–117
  longevity of, 4
  production inspection, 90
  raw, 5–6
  selling, 125
  spring, 105
  uncapped, 119–120
  uses for, 126–127
Honey cells, 124
Honey flow, 71, 119
Honey sacs, 115
Huber, Francis, 35

I

Inspections
  checklists, 89, 96
  defined, 29
  first two months, 101
  frequency of, 83–85
  Langstroth hives, 93–94
  number of bees to expect,
    82
  preparing for, 86–87
  procedure, 89–92
  recordkeeping, 95–97,
    129
  timing of, 81–83
  top bar hives, 94
  Warre hives, 95
Italian honeybees, 54–55

L

Langstroth, Lorenzo, 35–36
Langstroth hives, 35–37, 43,
  70, 93–94
Larvae, 16, 90
Legal considerations, 9–10,
  42, 67, 125
Lemongrass oil, 75
Livestock, 3

M

Moisture, 41, 50, 112, 131

N

Nasanov gland, 75
Nectar flow, 55, 102, 119
Neighbors, 11–12, 31
Nosema disease, 133
Nucs, 13, 56–57, 74–75

O

Observations, 85–86, 129
Optic flow, 32
Orientation flights, 76,
  106–107

P

Packs, 13, 56, 73
Pesticides, 32, 60–61
Pests, 134–138
Pets, 12
Pheromones, 26, 75
Pollen, 3–4, 5, 6
Pollen baskets, 102
Pollination, 5, 6
Pools, 78
Poop, 107, 133–134
Predators, 23, 31, 138–140
Propolis, 3, 22–23, 91–92
Pupae, 24, 90

Q

Queen cells, 17
Queen excluders, 93
Queens, 12, 15–17, 26, 72–73, 83, 137, 145–146
Quilting boxes, 41

R

Rain, 103
Refractometers, 50, 119
Requeening, 137, 145–146
Robber bees, 139
Royal jelly, 3–4
Russian honeybees, 54–55

S

Sacbrood, 136
Scouts, 22. *See also* Worker bees
Sickness, in bees
    common pests, 134–138
    preventing, 130–131
    reasons for, 129–130
    signs of, 131–133
Small hive beetles, 138
Smokers, 47, 48, 49, 71, 87–89, 148
Spinning extraction method, 124
Spring, 102–105
Stings, 7, 97–99
Summer, 105–107
Supers, 130
Supersedure cells, 26
Supplies, 46–51, 120–121, 147–148
Swarm capture, 58
Swarm cells, 26
Swarms, 8, 104, 144–145

T

Top bar hives, 37–40, 43, 94
Tracheal mites, 137
Transportation, of bees, 66, 68

V

Varroa mites, 135–136
Veils, 47, 48, 147

W

Waggle dance, 6, 23
Warré, Émile, 41
Warre hives, 41–43, 95
Water dishes, 51, 79, 147
Wax moths, 137–138
Winter, 111–113
Worker bees, 6, 17–20, 25

ACKNOWLEDGMENTS

A SPECIAL THANKS to and deep appreciation for my two beekeeping mentors, Patricia Candal and Rick Duvall. Thank you for answering all the calls and texts, and coming to my rescue over the years every time I had a bee question or problem. I couldn't have done it without you.

My good buddy Sam Crandall, every homesteader should have a Sam. Beyond appreciative for all of your help. To Laurie Neverman for believing that I could.

ABOUT THE AUTHOR

 AMBER BRADSHAW loves foraging for her food, gardening, and cooking outdoors. Bees and beekeeping have been an essential part of her gardening and farmers' market success. Amber is a natural beekeeper and provides her bees with a safe, toxin-free environment. She teaches others how to become self-sufficient by making eco-friendly products. When she's not in her gardens, you can find her working online.

Being a former 4-H leader, author, public speaker, owner and operator of an online farmers' market, herb society president, and extended learning instructor at a local college, Amber loves living as close to nature as possible. She currently resides in the Smoky Mountains of east Tennessee with her family of five, where they filmed the construction of their off-grid home for the Discovery Channel.

She is happy to share her knowledge with others through public speaking events, private instruction, and online mentoring. To learn more, you can visit her website MyHomesteadLife.com.